THE LIBRARY
ST. MARY'S COLLEGE OF MARYLAND
ST. MARY'S CITY, MARYLAND 20686

THE FRONTIERS COLLECTION

THE FRONTIERS COLLECTION

Series Editors:
D. Dragoman M. Dragoman A.C. Elitzur M.P. Silverman J. Tuszynski H.D. Zeh

The books in this collection are devoted to challenging and open problems at the forefront of modern physics and related disciplines, including philosophical debates. In contrast to typical research monographs, however, they strive to present their topics in a manner accessible also to scientifically literate non-specialists wishing to gain insight into the deeper implications and fascinating questions involved. Taken as a whole, the series reflects the need for a fundamental and interdisciplinary approach to modern science. It is intended to encourage scientists in all areas to ponder over important and perhaps controversial issues beyond their own speciality. Extending from quantum physics and relativity to entropy, time and consciousness – the Frontiers Collection will inspire readers to push back the frontiers of their own knowledge.

Quantum Mechanics and Gravity
By M. Sachs

Mind, Matter and Quantum Mechanics
By H. Stapp

Quantum–Classical Correspondence
By A.O. Bolivar

Quantum–Classical Analogies
By D. Dragoman and M. Dragoman

Knowledge and the World:
Challenges Beyond the Science Wars
Edited by M. Carrier, J. Roggenhofer,
G. Küppers, and Ph. Blanchard

Life – As a Matter of Fat
By O.G. Mouritsen

Series homepage – springeronline.com

Ole G. Mouritsen

LIFE – AS A MATTER OF FAT

The Emerging Science of Lipidomics

Prof. Ole G. Mouritsen
University of Southern Denmark, MEMPHYS-Center for Biomembrane Physics,
Physics Department, Campusvej 55, 5230 Odense M, Denmark   email:ogm@memphys.sdu.dk

Series Editors:

Prof. Daniela Dragoman
University of Bucharest, Physics Faculty, Solid State Chair, PO Box MG-11,
76900 Bucharest, Romania   email: danieladragoman@yahoo.com

Prof. Mircea Dragoman
National Research and Development Institute in Microtechnology, PO Box 38-160,
023573 Bucharest, Romania   email: mircead@imt.ro

Prof. Avshalom C. Elitzur
Bar-Ilan University, Unit of Interdisciplinary Studies,
52900 Ramat-Gan, Israel   email: avshalom.elitzur@weizmann.ac.il

Prof. Mark P. Silverman
Department of Physics, Trinity College,
Hartford, CT 06106, USA   email: mark.silverman@trincoll.edu

Prof. Jack Tuszynski
University of Alberta, Department of Physics, Edmonton, AB,
T6G 2J1, Canada   email: jtus@phys.ualberta.ca

Prof. H. Dieter Zeh
University of Heidelberg, Institute of Theoretical Physics, Philosophenweg 19,
69120 Heidelberg, Germany   email: zeh@urz.uni-heidelberg.de

ISSN 1612-3018
ISBN 3-540-23248-6 Springer Berlin Heidelberg New York

Library of Congress Control Number: 2004115728

This work is subject to copyright. All rights are reserved, whether the whole or part of the material is concerned, specifically the rights of translation, reprinting, reuse of illustrations, recitation, broadcasting, reproduction on microfilm or in any other way, and storage in data banks. Duplication of this publication or parts thereof is permitted only under the provisions of the German Copyright Law of September 9, 1965, in its current version, and permission for use must always be obtained from Springer. Violations are liable to prosecution under the German Copyright Law.

Springer is a part of Springer Science+Business Media

springeronline.com

© Springer-Verlag Berlin Heidelberg 2005
Printed in Germany

The use of general descriptive names, registered names, trademarks, etc. in this publication does not imply, even in the absence of a specific statement, that such names are exempt from the relevant protective laws and regulations and therefore free for general use.

Typesetting by the author
Data conversion by TechBooks
Cover design by KünkelLopka, Werbeagentur GmbH, Heidelberg

Printed on acid-free paper   54/3141/jl - 5 4 3 2 1 0

To Myer Bloom, mentor in science and life

Acknowledgments

The research work underlying the picture of fats, lipids, and membranes advocated in the present book derives from many different scientists and students in several laboratories across the world. The specific examples and data presented are however biased toward the work carried out by the author and his collaborators over the last almost twenty years. Therefore, the book should not be considered an authoritative monograph but more a personal perspective on a diverse and rapidly expanding field of science. The work has been supported by a number of public and private funding agencies, notably the Danish Natural Science Research Council, the Danish Technical Research Council, the Danish Medical Research Council via the Center for Drug Design and Transport, the Villum Kann Rasmussen Foundation, the Carlsberg Foundation, the Velux Foundation, the Hasselblad Foundation, and the Danish National Research Foundation. The author has over the years benefited from stimulating interaction and fruitful collaboration with a large number of colleagues and students, in particular from the Center for Biomembrane Physics (MEMPHYS) and from the Canadian Institute for Advanced Research's program on the Science of Soft Surfaces and Interfaces under the directorship of Professor Myer Bloom. The author is greatly indebted to the following: Thomas Andresen, Luis Bagatolli, Rogert Bauer, Thomas Baekmark, Gerhand Besold, Rodney Biltonen, Thomas Bjørnholm, Myer Bloom, David Boal, Thomas Hønger Callisen, Robert Cantor, Bernd Dammann, Jesper Davidsen, Lars Duelund, Evan Evans, Sven Frøkjær, Tamir Gil, Henriette Gilhøj, Per Lyngs Hansen, Jonas Henriksen, Pernille Høyrup, John Hjort Ipsen, Ask Jacobsen, Morten Ø. Jensen, Claus Jeppesen, Kent Jørgensen, Thomas Kaasgaard, Danielle Keller, Lars Kildemark, Dennis Kim, Paavo Kinnunen, Beate Klösgen, Per Knudsen, Mohamed Laradji, Chad Leidy, Jesper Lemmich, Ling Miao, Kell Mortensen, Mohan Narla, David Needham, Morten Nielsen, Jaan Noolandi, Adrian Parsegian, Tina Pedersen, Günther Peters, Amy Rowat, Jens Risbo, Mads C. Sabra, Erich Sackmann, Adam C. Simonsen, Maria M. Sperotto, Jenifer Thewalt, Christa Trandum, Ilpo Vattulainen, Peter Westh, Matthias Weiss, Michael Wortis, and Martin Zuckermann. Olaf Sparre Andersen is thanked for collaboration concerning a conference book on which part of Chap. 7 is based. Michael Crawford shared with me his insight into the relationships between lipids, nutrition, brain evolution and human health. I am indebted to Rodney Cotterill for useful advise concerning the book title. Fabienne Piegay, Laurence Sideau, and Martin Bennetzen are thanked for thorough and critical reading of the manuscript.

I have drawn on the help of a number of colleagues and the permission to use their graphic material for preparing many of the figures of this book. A list of the sources for the figures is provided at the end. Tove Nyberg has kindly drawn many of the figures. Jonas Drotner Mouritsen is thanked for creative suggestions and helpful advice regarding the preparation and design of many of the illustrations.

The Villum Kann Rasmussen Foundation is gratefully acknowledged for a generous grant that made it financially possible to arrange for the leave of absence that was required to get free time to start working on the book. Institutionen San Cataldo, where the first version of the book manuscript emerged in the tranquillity of Cell No. 6 of the old convent San Cataldo overlooking the beautiful Amalfi Bay in Southern Italy, is thanked for its hospitality during the month of April 2001.

Contents

Prologue: Lipidomics – A Science Beyond Stamp Collection .. 1

Part I The Overlooked Molecules

1 **Life from Molecules**... 9
 1.1 The Three Kingdoms of Life 9
 1.2 The Molecules of Life 10
 1.3 The Post-Genomic Era 16
 1.4 A Call for Physics 19

2 **Head and Tail** 23
 2.1 Fat Family: Fats and Fatty Acids 23
 2.2 The Polar Lipids – Both Head and Tail 26
 2.3 Cholesterol – A Lipid of Its Own 28
 2.4 Strange Lipids 29
 2.5 Lipid Composition of Membranes 31

3 **Oil and Water** 33
 3.1 Water – The Biological Solvent 33
 3.2 The Hydrophobic Effect 34
 3.3 Mediating Oil and Water 35
 3.4 Self-Assembly and the Lipid Aggregate Family 37
 3.5 Plucking Lipids 40

4 **Lipids Speak the Language of Curvature** 43
 4.1 How Large Is a Lipid Molecule? 43
 4.2 Lipid Molecules Have Shape 46
 4.3 Lipid Structures with Curvature 46
 4.4 Microorganisms' Sense of Curvature 51

5 **A Matter of Softness** 53
 5.1 Soft Matter 53
 5.2 Soft Interfaces 53
 5.3 Forces Between Soft Interfaces 58

	5.4 Lipid Membranes are Really Soft 59
6	**Soft Shells Shape Up** 63
	6.1 Bending Interfaces 63
	6.2 Spontaneous Curvature 66
	6.3 Shaping Membranes 67
	6.4 Red Blood Cells Shape Up 68
7	**Biological Membranes – Models and Fashion** 73
	7.1 What Is a Model? 73
	7.2 Brief History of Membrane Models......................... 74
	7.3 Do We Need a New Membrane Model? 76
	7.4 Theoretical and Experimental Model Systems 78

Part II Lipids Make Sense

8	**Lipids in Bilayers – A Stressful and Busy Life** 81
	8.1 Trans-bilayer Structure..................................... 81
	8.2 The Lateral Pressure Profile 83
	8.3 How Thick Are Membranes? 85
	8.4 Lively Lipids on the Move 86
9	**The More We Are Together** 91
	9.1 Phase Transitions Between Order and Disorder 91
	9.2 Lipids Have Phase Transitions 94
	9.3 Mixing Different Lipids..................................... 98
	9.4 Cholesterol Brings Lipids to Order 100
10	**Lipids in Flatland**... 105
	10.1 Gases, Liquids, and Solids in Two Dimensions.............. 105
	10.2 Langmuir and Langmuir-Blodgett Films 107
	10.3 Pattern Formation in Lipid Monolayers 110
	10.4 Lipids Make the Lung Work 112
11	**Social Lipids** 117
	11.1 Lateral Membrane Structure 117
	11.2 Imaging Lipid Domains 121
	11.3 Lipid Rafting 124
	11.4 Domains and Rafts Carry Function 127

12	**Lively Lipids Provide for Function** 129
	12.1 Leaky and Thirsty Membranes 129
	12.2 Repelling Membranes 131
	12.3 Enzymes Can Sense Membrane Transitions 133
	12.4 Lipid Thermometer in Lizards 135

13	**Proteins at Lipid Mattresses** 137
	13.1 Coming to Terms with Lipids 137
	13.2 Anchoring at Membranes 139
	13.3 Spanning the Membrane 142

Part III Lipids in Action

14	**Cholesterol on the Scene** 149
	14.1 Molecule of the Century 149
	14.2 Evolutionary Perfection of a Small Molecule 150
	14.3 Cholesterol Fit for Life 152
	14.4 Cholesterol As a Killer 156

15	**Lipids in Charge** 159
	15.1 Lipids and Proteins Match Up 159
	15.2 Stressing Proteins to Function 165
	15.3 Lipids Opening Channels 169
	15.4 Lipids Mediate Fusion 170

16	**Being Smart – A Fishy Matter of Fat** 173
	16.1 The Essential Fatty Acids 173
	16.2 Evolution of the Human Brain 175
	16.3 Lipids at the Border of Madness 177

17	**Liquor and Drugs – As a Matter of Fat** 181
	17.1 Lipids Are Targets for Drugs 181
	17.2 Alcohol and Anesthesia 182
	17.3 Poking Holes in Membranes 184
	17.4 Gramicidin – the Portable Hole 186

18	**Lipid Eaters** 189
	18.1 Enzymes That Break Down Lipids in Crowds 189
	18.2 Watching Enzymes at Work 192
	18.3 Lipids Going Rancid 193

19 Powerful and Strange Lipids at Work 197
  19.1 The Impermeable Barrier – Lipids in the Skin 197
  19.2 Surviving at Deep Sea and in Hot Springs 200
  19.3 Lipids As Messengers 203
  19.4 Lipids As a Matter of Death 205

20 Survival by Lipids 209
  20.1 Lipids for Smart Nanotechnology 209
  20.2 Lipids Deliver Drugs 214
  20.3 Liposomes As Magic Bullets 216
  20.4 Lipids Fighting Cancer 219

Epilogue: Fat for Future 225

Bibliography 227

Sources for Figures 249

Index 255

Abbreviations

Fatty acids
AA   arachidonic acid
DHA   docosahexaenoic acid
DPA   docosapentaenoic acid
EPA   eicosapentaenoic acid

Lipid polar head groups
DGDG   digalactosyl diglyceride
PC   phosphatidylcholine
PE   phosphatidylethanolamine
PG   phosphatidylglycerol
PI   phosphatidylinositol
PS   phosphatidylserine

Lipids
DAG   di-acylglycerol
DAPC   di-arachioyl PC
DCPC   di-decanoyl PC
DLPC   di-laureoyl PC
DMPC   di-myristoyl PC
DMPE   di-myristoyl PE
DMPG   di-myristoyl PG
DOPC   di-oleoyl PC
DPPC   di-palmitoyl PC
DPPE   di-palmitoyl PE
DSPC   di-stearoyl PC
DSPE   di-stearoyl PE
POPC   palmitoyl-oleoyl PC
POPE   palmitoyl-oleoyl PE
SM   sphingomyelin
SOPC   stearoyl-oleoyl PC

Others
ATP   adenosine triphosphate
DNA   deoxyribonucleic acid
ER   endoplasmic reticulum
HDL   high-density lipoprotein
LDL   low-density lipoprotein
RNA   ribonucleic acid
ROS   reactive oxygen species

Prologue:
Lipidomics – A Science Beyond Stamp Collection

To most people, fats are something vicious that are dangerous to our health and well-being and therefore should be avoided. Some people know that fats are essential ingredients of the diet and, furthermore, provide for tasty meals. Others acknowledge that certain fats like cholesterol are required for the body's production of important hormones, such as sex hormones, as well as vitamin D and the bile you need in your stomach to break down the food. Many of us appreciate that the unsaturated fats present in seafood provide for good health and longevity. However, few realize that fats are as important for life as proteins and genes. And probably very few people know that in terms of mass, fat is the most important part of our brain and the second most important of all other soft tissue.

In the sciences, fats are called lipids. Lipids are studied by nutritionists who investigate how the intake of fats in our diet affects the composition of various parts of the body, e.g., the heart, the liver, and the brain. You are what you eat! Lipids are also studied by biochemists who investigate the synthesis and breakdown of lipids as a source of energy and building material for cells. Among molecular biologists and physical scientists, lipids are less appreciated, although lipids are listed along with proteins, genes (DNA and RNA), and carbohydrates (sugars) as the fundamental building blocks of all living matter. Proteins and genes are known to be very specific to the functions they perform. The same is true for the carbohydrates that are used by the living cell to recognize foreign substances and to identify other cells.

Lipids as structure builders and fat depots are, however, most often characterized by terms like variability, diversity, plasticity, and adaptability. Hence, lipids appear to play a fairly nonspecific role, being rather dull and anonymous compared to fashionable stuff like the proteins that catalyze all biochemical reactions and the genes that contain the information needed to produce the proteins. There are no genes coding for lipids as such, only for the enzymes that build and modify the lipids which, for example, we obtain from our food.

There is a particular reason why lipids are considered dull and less interesting. This reason has to do with the powerful concept of molecular *structure* that has permeated life scientists' way of thinking throughout the twentieth century. The central dogma is that molecular structure controls function.

Lipids apparently do not have the intriguing molecular structures which important molecules like proteins and DNA have. Moreover, lipids do not as easily as proteins and DNA lend themselves to revealing the secrets of how molecular structure leads to biological function. In fact, lipids are often tacitly assumed to constitute a structureless fatty material which at best is organized in a membrane structure that plays the role of a passive container of the cell and an appropriate solvent or template for the important molecules of life: Lipids grease the functional machinery of the cells controlled and run by proteins and DNA.

The traditional view of lipids as dull molecules has changed considerably over the last decade, and their importance for cell function and health is becoming more recognized. This is what the present book is about. Lipids are proposed to be as important for life as proteins and genes. The book will point out that lipids lead to interesting and intriguing structures with very unusual and subtle materials properties that have been optimized by evolutionary principles over billions of years. These properties are consequences of the fundamental physical principles of self-organization that rule when many molecules act in concert. A key player in this concert is *water*, which functions as the unique biological solvent. Water is omnipresent in all functioning biological systems. The peculiar properties of water force lipid molecules to self-assemble and organize into subtle structures. In particular, lipids in water can form lipid bilayers and *membranes*. Lipid membranes are extended thin layers that are only two molecules thick. These layers constitute the backbone of all biological membranes. Membranes are an ubiquitous structural element in all living cells.

And cells are literally packed with membranes. In a human being, which is composed of about 10^{14} cells, the total surface of the membranes has been estimated to cover an area of about $100 \, \text{km}^2$. An important membrane function is purely topological: Membranes compartmentalize living matter into cells and subcellular structures. Furthermore, membranes present themselves to macromolecules as highly structured interfaces at which important biochemical processes are carried out and catalyzed. Obviously, the structure and molecular organization of the lipid-bilayer component of membranes hold the key to understanding the functioning of membranes.

Due to the fact that lipids form membranes by self-assembly processes that do not involve strong chemical forces, membranes are pieces of *soft matter*. Softness is a materials feature that lipid membranes share with other forms of condensed matter like polymers and liquid crystals. During evolution, nature has evolved biological membrane structures as an optimal form of micro-encapsulation technology that, on the one hand, imparts the necessary durability to the particular soft condensed matter that membranes are made of and, on the other hand, sustains the lively dynamics that are needed to support and control the mechanisms of the many essential cellular functions associated with membranes.

Lipids are now known not only to be structural builders of the cell and an energy source for the cell functioning. In the form of membranes, lipids are crucial for controlling indirectly a great variety of biological functions that take place at or are mediated by membranes. Evidence emerging from recent research has shown that the functional role of membrane lipids may be as important as that of proteins. The lipids not only act as a passive solvent for the proteins and as a means of compartmentalization, but are also an integral part of cellular function. Many lipid species are now known to play a very active role, serving as so-called second messengers that pass on signals and information in the cell. Lipids also play the roles of enzymes, receptors, drugs, as well as regulators of, e.g., neurotransmitter activity. Lipids are known to modulate the expression of genes. Disorder in the lipid spectrum of cells has recently been related not only to atherosclerosis but also to major psychiatric illnesses. Finally, poly-unsaturated lipids via the diet have been proposed to provide an evolutionary driving force supplementing Darwinian natural selection. All these factors emphasize a necessary shift from a one-eyed genocentric approach to membrane science toward a more balanced organocentric view. This puts the study of the physical properties of membranes at a very central position.

We have just witnessed a scientific revolution that at the end of the twentieth century took us through the *genomics* era with focus on sequencing genes and mapping gene products. We are now in the middle of the *proteomics* era in which the multitude of proteins that are coded for by the genes are being identified and their functions unveiled. In front of us we have a new era of recognizing the role of lipids. A new science termed *lipidomics* is emerging. In a not-too-distant future, we can imagine all the "omics" to converge into the grand *metabolomics* era in which the interplay between all the constituents of living matter is being studied. This development mirrors a gradual anticipation of the actual complexity of the problem.

Some have claimed that life sciences, due to the strong focus on sequencing and the mapping of genes and proteins, have been turned into a data-driven information science. It is certainly true that modern biology has for ever been changed by the tremendous wealth of information that has been derived from the genomics and proteomics programs. However, a deep insight into the workings of living systems requires more than collection of data and information processing. Science is by nature driven by hypotheses, and it goes beyond stamp collection. This is certainly also true of lipidomics. A mapping of all lipid species in all cell types is undoubtedly useful. However, a new lipidomics science has to go beyond that and approach cellular functioning from a more holistic perspective. And it will have to be hypothesis-driven. Lipidomics must involve a quantitative experimental and theoretical study of, e.g., lipid and membrane self-assembly, lipid-protein interactions, lipid-gene interactions, and the biophysical properties of lipid structure and dynamics.

The term lipidomics has only appeared in the scientific literature within the last few years. The name of this most recent member of the 'omics' family seems to have been brought forward and discussed at a number of international conferences during 2001 and was first suggested in the scientific literature by Leif Rilfors and Gøran Lindblom in 2002 in the context of *functional lipidomics*. A debate at the 2002 congress of the International Society for the Study of Fatty Acids and Lipids decided to use the term lipidomics in much the same manner as it is being used in the present book, that is in a way that goes considerably beyond mapping out all lipids and their functions in a traditional systems-biology fashion.

A deeper knowledge of the physical properties of lipids is essential for understanding the living world and its modes of functioning. It may also be useful in revealing the causes of malfunctioning and diseased conditions and how such malfunctioning can be restored by medical treatment or by altered living conditions. The insights into the physics and the inner workings of the cellular machinery, including the role of lipids, can furthermore be lessons for technology. Examples include the development of futuristic soft and biocompatible materials, functionalized surfaces and sensors, the design of new and effective drugs and drug-delivery systems for, e.g., cancer therapy, as well as the design of new enzymes for the cheap and clean production of drugs and chemicals.

This book presents a personal and multidisciplinary perspective on the physics of life and the particular role played by lipids and the lipid-bilayer component of cell membranes. The emphasis is on the physical properties of the lipid membrane seen as a soft and molecularly structured interface. By combining and synthesizing insights obtained from a variety of recent studies, an attempt is made to clarify what membrane structure is and how it can be quantitatively described. Furthermore, it will be shown how lipid membrane structure and organization can control functional properties of membranes. The strategy of the book is to provide a bridge between, on the one side, the microscopic world of membranes, i.e., the world of the molecules, and, on the other side, the macroscopic world, i.e., the world as we observe and sense it. This involves unravelling the organizational principles that govern the many types of structure that arise on length scales from the size of the individual molecule, across molecular assemblies of proteins and lipid domains in the range of nanometers, to the meso- and macroscopics of whole cells.

The book provides a Bibliography containing a selected list of references to other books, review papers, and research articles accounting for most of the factual statements made in the book. These references to the literature have been selected according to a minimal principle. The rule has been adopted that references are made to recent publications and not necessarily to the original work. From the references given, the interested reader should be able to track down the original literature. The reader is furthermore referred to the list of specialized books and review papers for a more comprehensive list of references. I apologize in advance to those authors and colleagues who may feel that their original work should have been referenced and discussed in more detail.

A Note on Length Scales, Forces, Energy, and Temperature

The picture of lipids described in this book takes its starting point in the molecular world where the molecules move around due to influence of temperature. It is hence useful to describe the various entities on scales described in units of nanometers. One nanometer (nm) is 10^{-9} m, i.e., one millionth of a millmeter. Cell sizes are conveniently given in units of micrometers. One micrometer (μm) is 1,000 nm. The appropriate unit of force for biological molecules on the nm scale is pico-Newton (pN). One pN is 10^{-12} N. These units sound terribly small. It is helpful to combine them in the form of the energy that corresponds to the thermal energy at room temperature ($T_{room} = 293K = 20°C$)

$$\text{Thermal energy} = k_B T_{room} = 4.1 \cdot \text{nm} \cdot \text{pN}, \tag{1}$$

where k_B is Boltzmann's constant, which is a universal number. Most of the phenomena characteristic of biomolecular structure and dynamics are strongly influenced by temperature, and it is therefore of importance whether the involved energies are smaller or larger than the thermal energy in (1). As a rule of thumb one would say that if a characteristic energy of some association, e.g., a binding between two molecular-scale objects, is of the order of a few $k_B T_{room}$ or less, thermal agitation should be significant, and the lifetime of the association can be short. A couple of examples can serve as an illustration. A covalent chemical bond C–C between two carbon atoms represents an energy of the order of $100 k_B T_{room}$ (with a force equivalent of around 5,000 pN) and is therefore very stable at room temperature. A typical hydrogen bond amounts to about $10 k_B T_{room}$ (with a force equivalent of around 200 pN) and is therefore often influenced by thermal agitation. Turning then to weak physical interactions, the van der Waals interaction between two methane molecules, or other hydrocarbon moieties, represents about $1 k_B T_{room}$ (with a force equivalent of around 40 pN). Even lower energies and correspondingly lower forces govern the weak molecular associations in biological systems, e.g., the binding of a small enzyme to a membrane surface amounts to about 20 pN, the force required to pull out a single lipid molecule of a membrane is only about 2 pN, and finally the motor proteins that function, e.g., in muscle contraction, exert forces as small as 1 pN. In all these cases, thermal agitation is of major importance.

Part I

The Overlooked Molecules

1 Life from Molecules

1.1 The Three Kingdoms of Life

Living organisms are divided into three kingdoms: the *eukaryotes*, the *eubacteria*, and the *archaebacteria*. The eubacteria and the archaebacteria, which among themselves differ as much as they do from eukaryotes, are conventionally grouped together as *prokaryotes*. The bacteria common to most people, e.g., coli bacteria or the bacteria in sour milk, are eubacteria. Archaebacteria are typically found in rather hostile environments, such as in hot springs, at the bottom of deep sea, or in the very acidic milieu of the cow's stomach. These bacteria do not tolerate oxygen and they often present a health hazard to humans. Eukaryotes are animals, plants, and fungi and also include single-cell organisms like yeast. Figure 1.1 gives examples of single cells from the three kingdoms.

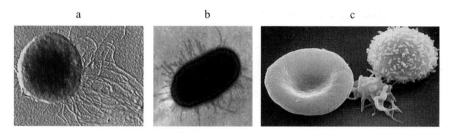

Fig. 1.1a–c. Examples of cells from the three kingdoms of life. (**a**) An archaebacterium: *Methanococcus jannischiiwas*. Diameter about 2 µm. (**b**) A eubacterium: *Escherichia coli*. Size about 2–3 µm. (**c**) Eukaryotes: human red and white blood cells shown together with a platelet. The diameter of the red blood cell is about 6 µm. The cells from the three kingdoms are not drawn on the same scale

Despite their difference in appearance and functioning, archaebacteria, eubacteria, and eukaryotes are all made from the same basic molecular building blocks, and they are all based on the same chemistry. Although it is generally believed that all cells have a common ancestor, the ancient evolutionary history of the different cell types is subject to considerable dispute. The three

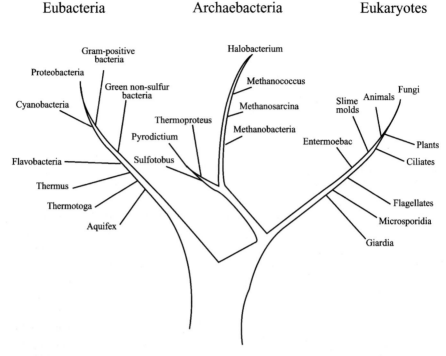

Fig. 1.2. Phylogenetic tree with the three kingdoms of life: eubacteria, archaebacteria, and eukaryotes. The relative distance between the organisms is proportional to the evolutionary distance as determined by ribosomal-RNA nucleotide sequencing

kingdoms and their interrelations are often represented by a so-called universal *phylogenetic tree,* as shown in Fig. 1.2.

Although the origin of life on Earth is a controversial and unresolved problem, it is a reasonable assumption that the first cellular living systems on Earth were assembled from four types of molecular building blocks: (i) information-storing molecules capable of reproduction, (ii) enzyme-like catalysts encoded by that information and able to enhance reproduction rates, (iii) molecules capable of storing energy and using this energy to convert molecules into organized assemblies of biologically active molecules, and (iv) special boundary-forming molecules capable of encapsulating and protecting the former three types of molecules. The last category of molecules is the focus of the present book.

1.2 The Molecules of Life

All cells are built from small organic molecules that are based on the chemistry of carbon. These small molecules belong to essentially four classes: the

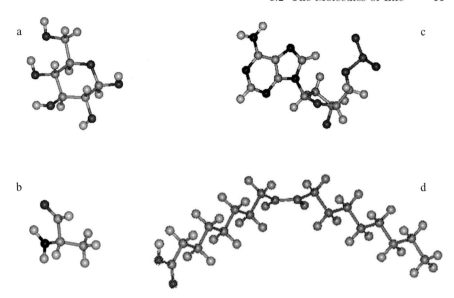

Fig. 1.3a–d. Examples representing the four classes of small organic molecules that are the building blocks of all living matter. (**a**) Sugar: glucose. (**b**) Amino acid: alanine. (**c**) Nucleotide: adenosine. (**d**) Fatty acid: oleic acid

sugars, the *amino acids*, the *nucleotides*, and the *fatty acids*. Examples of these small elementary building blocks of living matter are given in Fig. 1.3.

The small organic molecules are combined with other molecules from the same class or with molecules from the other classes to make larger entities, so-called macromolecules or macromolecular assemblies. There are basically four classes of these larger entities, the *polysaccharides*, the *proteins*, the *nucleic acids*, and the *fats* (lipids and membranes), as illustrated in Fig. 1.4.

Proteins are also called poly-amino acids (or *polypeptides*), and nucleic acids are called *polynucleotides*, reflecting the fact that proteins and nucleic acids, just like polysaccharides, are biopolymers, i.e., long-chain molecules composed of many monomers that are bound together by strong chemical bonds.

Since there are about twenty different types of amino acids in nature and since a protein can consist of up to several hundred amino acids, a very large number of different proteins can be perceived. Similarly, the five different nucleotides used by nature allow for an immense richness in different nucleic acids that make up DNA (deoxyribonucleic acids) and RNA (ribonucleic acids). Like proteins, the nucleic acids are linear molecules, and it is the particular sequence of the monomers that determines the properties of both proteins and nucleic acids. DNA and RNA contain the genetic information that is organized in *genes*. The entire DNA string of an organism is termed the *genome*, which can contain millions of nucleotides. For example,

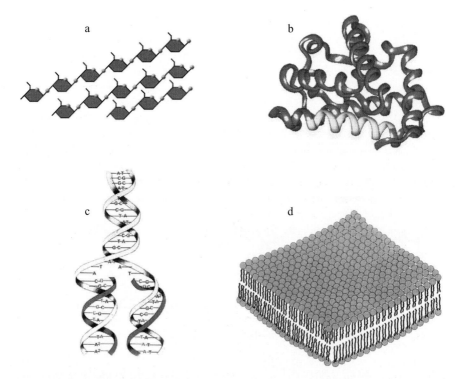

Fig. 1.4a–d. Examples representing the four classes of larger macromolecular entities of which all living matter is composed. (**a**) Polysaccharide: cellulose. (**b**) Protein: myoglobin. (**c**) Nucleic acid: DNA in the form of a double helix. (**d**) Lipid assembly: lipid bilayer membrane. The different macromolecules and assemblies are not drawn on the same scale

the human genome includes about 20,000–25,000 genes composed of a little less than three billion nucleotides. In addition to encoding genetic information, nucleotides also perform functions as energy carriers (ATP, adenosine triphosphate), catalysts, and messengers.

When it comes to the sugars, living organisms exploit a large number of different monosaccharides. Hence, it is not uncommon to find hundreds of different polysaccharides in a cell. Sugars allow for additional complexity in the type of materials that can be built because they can combine into branched macromolecular networks. Such networks are responsible for forming biological fibers and scaffolding and are also an important part of the cell's recognition system.

We are then left with the fatty acids and the lipids. In contrast to the sugars, the amino acids, and the nucleotides, lipids do not link chemically to form "poly-lipids." No such thing exists under natural conditions. Instead, they form "loose" macromolecular (or supramolecular) assemblies, of which the *lipid bilayer membrane*, shown in Fig. 1.4d, is the most prominent example. In some cases, lipids combine chemically with proteins and sugars. However, when forming living matter, lipids usually maintain their molecular integrity. The lipid bilayer is the core of all biological membranes. A lipid bilayer membrane contains billions of lipid molecules and a cell membrane often contains hundreds of different kinds of lipids.

All cells of living beings are confined and compartmentalized by a number of membranes, as illustrated in Fig. 1.5 in the case of eukaryotes. Common for all cells is a cell-surface membrane called the *plasma membrane*. The plasma membrane is a very stratified and composite structure whose central element is the lipid bilayer, as illustrated in Fig. 1.6 in the case of a very simple unicellular microorganism, *Escherichia coli*. The lipid bilayer is extremely thin in comparison to the size of the cell it encapsulates. A schematic cartoon of the plasma membrane of a eukaryotic cell with all the other components

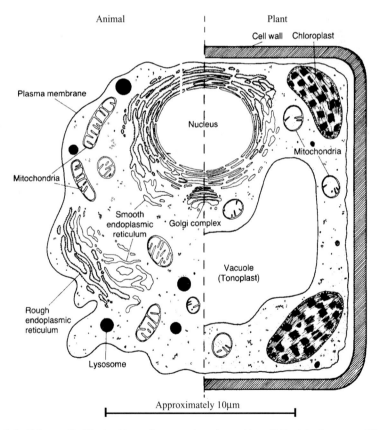

Fig. 1.5. Schematic illustration of a generic eukaryotic cell that is drawn artificially to compare an animal cell (*left*) and a plant cell (*right*). Plant cells, as well as bacterial cells, have an additional outer cell wall

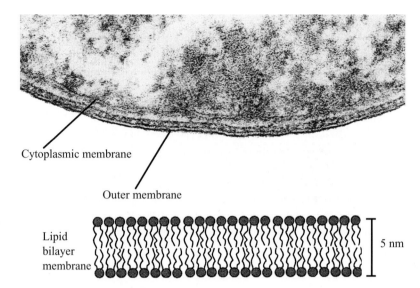

Fig. 1.6. Electron microscopy picture of the bacterium *Escherichia coli*. This Gram-negative species is encapsulated in an inner bilayer cytoplasmic membrane, an intermediate peptidoglycan layer, and an outer bilayer membrane. Also shown is a schematic illustration of the 5 nm thick lipid bilayer, which is the core of the membrane, as illustrated in more detail in Fig. 1.7

it contains in addition to the lipid bilayer is shown in Fig. 1.7. The plasma membrane is a unique composite of all the types of macromolecules described above except nucleic acids. Its molecular composition depends on the type of cell. Carbohydrates are a minor component with less than 10% of the dry mass. The weight ratio of proteins and lipids can vary from 1:5 to 5:1.

Whereas prokaryotic cells only have a plasma membrane and some less structured internal membrane systems, the eukaryotic cells have in addition a number of well-defined internal membranes associated with the cell *nucleus* and the *organelles* (cf. Fig. 1.5). The cell nucleus, which contains the cell's genetic material, is wrapped in a porous double membrane (the nuclear envelope). This membrane is topologically connected to the membranes of the *endoplasmic reticulum* (ER), which is the major site of synthesis of lipids and proteins. The *Golgi apparatus* contains a very convoluted agglomerate of membranes. This is where the newly synthesized molecules are modified, sorted, and packaged for transport to other organelles or for export out of the cell. The membranes of both Golgi and the ER are morphologically very complex and exhibit substantial curvature. The *mitochondria*, which contain their own DNA and RNA and produce their own proteins, contain two intertwined membranes, an inner and an outer membrane. The outer membrane acts as a sieve, retaining the larger proteins within its compartment. *Lysosomes* are rather small organelles bound by a membrane. The lysosomes operate as the

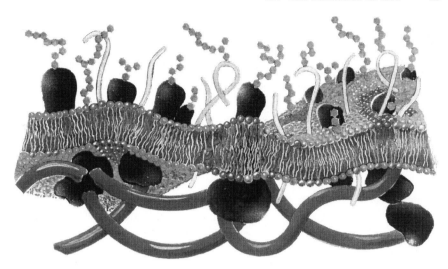

Fig. 1.7. Schematic model of the plasma membrane of a eukaryotic cell that highlights the membrane as a composite of a central lipid bilayer sandwiched between the carbohydrate glycocalyx (which consist of polysaccharides) on the outside and the rubber-like cytoskeleton (which is a polymeric protein network) on the inside. Intercalated in the lipid bilayer are shown various integral proteins and polypeptides. The membrane is subject to undulations, and the lipid bilayer displays lateral heterogeneity, lipid domain formation, and thickness variations close to the integral proteins. Whereas the lipid molecules in this representation are given with some structural details, the membrane-associated proteins remain fairly featureless. In order to capture many different features in the same illustration, the different membrane components are not drawn to scale

cell's garbage and recycling system, performing digestion, degradation, and export of unwanted molecules. Finally, there is a bunch of vesicles that support the extended trafficking needed by the cell to transport material within the cell and across the plasma membrane.

Membranes thereby become the most abundant cellular structure in all living matter. They can be considered nature's preferred mode of microencapsulation technology, developed as a means of compartmentalizing living matter and protecting the genetic material. The biological membrane is the essential capsule of life. Many important biological processes in the cell either take place at membranes or are mediated by membranes, such as transport, growth, neural function, immunological response, signalling, and enzymatic activity. An important function of the lipid bilayer is to act as a passive permeability barrier to ions and other molecular substances and leave the transmembrane transport to active carriers and channels.

Judging naively from Fig. 1.7, the role of lipids is much less glorious than that played by proteins and nucleic acids. Whereas proteins possess a specific molecular structure and *order* supporting function, and DNA has a very distinct molecular structure and order encoding the genetic information, lipids appear to be characterized by *disorder* and a lack of any obvious structural elements. The lipid bilayer is a molecular mess and it is hard to imagine any structural order in and among the lipids that is specific enough to control a delicate biological function.

This is the reason why lipids were not among the favorite molecules of the twentieth century's molecular and structural biologists. The lipids became the most overlooked molecules in biology.

In many ways, this is a paradox. Without comparison, lipids are the most diverse class of molecules in cells. Prokaryotic cells typically contain a hundred different types of lipids, and the larger eukaryotes cells many hundred different kinds. Moreover, the results of genomic research have revealed that more than 30% of the genome codes for proteins embedded in membranes. A possibly even larger percentage codes for proteins that are peripherally attached to membranes. If one adds to this the observation that the membrane-spanning parts of the proteins contain some of the most evolutionarily conserved amino-acid sequences, it becomes clear that lipids are indeed very important molecules for life. This insight calls for a deeper understanding of how membrane proteins and their function are related to the properties of the lipid bilayer membrane.

In order to understand the implications of disregarding lipids in the study of the physics of life and which challenges it has left us with at the beginning of the new millennium, we have to make a status on life sciences upon the entry to what has been called the *post-genomic* era. This era is characterized by an almost complete knowledge of genomes of an increasing number of different species ranging from bacteria, fungi, insects, worms, and mammals, including mouse and man.

1.3 The Post-Genomic Era

Molecular and structural biology have been some of the most successful sciences of the twentieth century. By focusing on the concept of *structure* – from the genes to the workhorses of living beings, the proteins – first revealing the genetic code, then the structure of many proteins, and very recently the whole genome of several species, including that of man, these sciences have had an enormous impact on life sciences and society. Structure, in particular, well-defined atomistic-level molecular structure, has been the lodestar in the quest for unravelling the genetic code and the properties of DNA, for understanding transcription of the code into protein synthesis, and for determining the properties of the proteins themselves. The relationship between macromolecular structure and function is simply the key issue in modern biology. The human genome project is probably the most monumental manifestation of the conviction among life scientists that gene structure is the Holy Grail of life.

1.3 The Post-Genomic Era

Knowing the genome of an organism implies information about which proteins the cells of this organism can produce. The genome is so to speak the blueprint of the proteins. Since we know how information is passed on from the genome to the proteins, we can also unravel the relationships between possible defects in a protein on the one side and errors and modifications in the genome on the other side. Such errors can lead to genetically determined diseases, such as cystic fibrosis. This insight can be of use in gene-therapeutic treatment of serious diseases, as well as in the production of plants and animals with desirable properties, e.g., plants that can better cope with poor weather conditions or are resistant to the attack of insects and microorganisms. Results of the genome research can also be used technologically to alter the genes of microorganisms like bacteria. Gene-modified bacteria can be exploited to produce useful chemicals and drugs.

Knowing the genome of an organism does not imply that one necessarily knows which function a given protein can carry out. It is not to be read in the genome why and how proteins carry out their various tasks, not even in the case where their function is known. Furthermore, it is not written in the genome how a cell and its various parts are assembled from the molecular building blocks. Neither can one read in the genome how biological activity is regulated or how cells are organized to become multi-cellular organisms of specific form and function. As a striking example, one cannot read in the genome why our fingers are almost equally long, how the leopard gets its spots, or what determines the width of the zebra's stripes.

The information contained in the genome is in this sense not complete, and additional principles have to be invoked in order to describe and understand the complex organization of the molecules of life. Complexity in living organisms does not come from the genome alone. One way of expressing this fact is to say that the genome provides the limitations and the space within which the biology can unfold itself. Biological function and pattern formation are *emergent phenomena* that arise in this space. This is the point where physics and the physics of complex systems come in. Physics is the generic discipline that in principle has the tools to predict and describe the emergent properties that are the consequences of the fact that many molecules are interacting with each other. The grand challenge in the post-genomic era consists of formulating and completing a program that combines results of genomic research with basic physics in a sort of *biophysical genomics* combined with proteomics and lipidomics.

The tremendous complexity of the problem we are facing perhaps becomes obvious when it is considered that the 30,000 or more genes in humans code for millions of different proteins. Each protein in turn can be in several different molecular conformations each of which may have its specific function. Moreover, many of the proteins often become post-translationally modified, e.g., installed with hydrocarbon chains. In addition to this, the functioning of a protein is modulated by its environment and how this environment is structured in space and time.

These challenges imply the provision of hypothesis-driven paradigms for understanding how cellular and subcellular structures of enormous complexity are formed out of their molecular building blocks, and how living systems are organized, regulated, and ultimately function. The old problem of bridging the gap between the genotype and phenotype still remains: Complete knowledge of a genome does not alone permit predictions about the supramolecular organization and functioning of a complex biological system.

Solving this problem is intellectually far more difficult than determining the genome of a species. For this purpose, principles from fundamental physics and chemistry are needed. Moreover, these principles will have to be developed and explored in a truly multidisciplinary setting. However, if this can be achieved, the post-genomic era will not only furnish the greatest challenges but also comprise some of the largest opportunities.

Biological membranes are outstanding examples of molecular assemblies of extreme complexity whose structure and function cannot be determined from the genome alone and which present some grand challenges to science. Due to the immense importance for life processes and not least the well-being of human beings, the study of biological membranes has for a long time been a central and very active field of research within medicine and biochemistry. Scientific disciplines like physiology, pharmacology, molecular biology, and nutritional science have all contributed to our current knowledge about biological membranes and their functions. However, the progress in the fundamental understanding of membranes has not been impressive compared to that related to proteins and DNA.

It is somewhat paradoxical that the preoccupation with well-defined molecular structure, which has led to so many successes in structural biology, may be the reason why an advance in the understanding of lipid membrane structure, and structure-function relationships for membranes, has been rather slow. The problem is that if one searches for well-defined structure in membranes in the same way as investigations are made of the structure of genes and proteins, one is going to utterly fail. The reason for this is that membranes are self-assembled molecular aggregates in which subtle elements of structure arise out of a state of substantial disorder, and where *entropy* consequently plays a major role. Disordered and partly ordered systems are notoriously difficult to characterize quantitatively. The challenge is to ask the right questions and to identify the hidden elements of order.

Although membranes consist of molecules (lipids, proteins, carbohydrates) of a well-defined chemical structure that are coded for in the genes, these molecules organize among themselves by physical principles that are nowhere to be found in the structure of the genetic material. The big question is then what these principles are and how they can operate to produce the robustness and specificity necessary for biological function. It is thought-provoking that the lipids in the biological membrane are not linked by strong and specific chemical forces in contrast to the amino acids in proteins and the nucleotides in DNA. Instead, they are kept together by weak and nonspecific physical forces, which we shall return to in Chap. 3. It is striking that nature has used a technology based on self-assembly processes in the construction of the essential capsule of all known life forms. Related to this question is the big mystery of lipid diversity. Why is it that membranes are composed of such large numbers of different lipid species?

As we shall see in Chap. 5, membranes seen as physical states of matter are fluid and soft interfaces, and they possess all the subtle structures of liquids and liquid crystals. Elucidation of the structure of membranes therefore requires concepts from the physics and physical chemistry of disordered materials and soft condensed matter. An increased understanding of the subtle physical properties of membranes viewed as soft biological materials is likely to lead to new insights as well as surprises. This insight is a prerequisite in the post-genomic era for effectively exploiting the wealth of structural information that becomes available. The goal is to understand the regulation of entire systems of cell organelles and whole cells, which involve complex, dynamic, and self-organized structures of membranes, biological fibers, and macromolecules that are constantly being transported, translated, and inserted into various parts of the cell.

1.4 A Call for Physics

The study of the physics of membranes is not an easy one and requires challenging experimental and theoretical approaches. Several circumstances have in recent years stimulated an interest in the physics of biological membranes.

Firstly, modern experimental techniques have provided quantitative information about the physical properties of well-defined model-membrane systems, seen as large self-organized assemblies of interacting molecules. This information has shown that the properties of membranes and aspects of their *biological function* are controlled by basic *physical principles*. Revealing and understanding these principles, along with a clarification of the nature of the feedback mechanism between physical properties and function, open up for rational ways of manipulating membrane function and malfunction.

Secondly, physicists and physical chemists have realized that biological systems, in particular membranes and proteins, are interesting objects of study in their own right: Membranes are structured and functional materials (soft interfaces) with unique material properties that are designed by nature during evolutionary times over billions of years. These natural materials are therefore in most respects functionally superior to man-made materials. In particular, natural materials are designed to be mechanically stable and to function on small scales (from nanometers to micrometers) and are therefore promising candidates for a whole new generation of *micro- and nanotechnology*. One example of an important biomedical application of membrane systems is the use of liposomes as biocompatible microcapsules in targeted drug delivery and gene therapy. Another example includes biosensors and medical micro-devices composed of immobilized enzymes or proteins attached to supported lipid membrane interfaces. We shall return to the various technological applications of lipids in Chap. 20.

In an attempt to understand, in molecular detail, how the functioning of biological membranes is related to their physical properties on different time and length scales, and how this relationship may be influenced by pharmaceutical drugs and environmental conditions, it is essential to characterize different membrane systems by means of a variety of powerful experimental and theoretical physical techniques.

In particular, it is necessary to achieve knowledge about the lateral structure and molecular organization of lipid membranes on length scales that are relevant to the particular membrane phenomenon in question. This puts focus on the nanometer scale and makes membrane science a truly nano-science. In fact, biological membranes as a micro-encapsulation technology can be seen as nature's preferred nanotechnology. Membrane science is concerned with an object that is 5 nm thick and has delicate structural features over scales from 1–1,000 nm. Indeed, biological membranes are optimized by evolutionary processes to function on the nanometer scale.

This perspective should be kept in mind when research strategies are chosen to investigate membranes and membrane models. Whereas many biological systems are accurately characterized on the molecular and atomic scale, as well on the large, cellular, and super-cellular level, there is a gap of knowledge at the intermediate subcellular scales that constitutes precisely the nanometer regime. It is in this regime where the workings of the complex cellular machinery is manifested. The advent in recent years of powerful theoretical methods and novel experimental techniques based on physical principles has opened a window to the nanometer world that calls for a renewed extensive study of membranes.

The nanoscopic organization of membranes is a key factor in various biological events governing the binding of molecules to the membrane, penetration and permeation of peptides and drugs, as well as insertion of membrane proteins. Moreover, the lateral structure controls the mechanical properties of the membrane and thereby its interaction with other membranes. The mechanical properties in turn are of crucial importance for the shape of cells, for cell cytosis and fusion, as well as for cell motility. In order to understand how proteins and enzymes function in membranes, e.g., in relation to transport, biochemical signaling, energy transduction, receptor-ligand interactions, and nerve activity, it is necessary to determine the ways in which proteins interact with the lipid bilayer, specifically how the proteins influence the local structure and composition of the bilayer, on the one hand, and how changes in the lipid-bilayer physical properties modulate the functional state of the proteins on the other hand.

Answering these questions is the challenge to the emerging science of lipidomics, and the answers hold the key to understanding fundamental aspects of the nanoscopic "machinery" of the cell.

2 Head and Tail

2.1 Fat Family: Fats and Fatty Acids

Oils and fats refer to a large and diverse group of chemical compounds that do not easily dissolve in water. There is no strict distinction between oils and fats; fats usually refer to materials like wax, lard, and butter that are solid at room temperature, whereas oils like olive oil and fish oil are liquid. As is well known, butter can melt upon heating and olive oil solidify by freezing. Fats are just frozen oils.

The main part of a fat or an oil is a hydrocarbon moiety, typically a *long-chain hydrocarbon*, as shown in Fig. 2.1. Hydrocarbon chains can contain different numbers of carbon atoms, and the bonds between the carbon atoms can be single bonds (*saturated*) or double bonds (*unsaturated*). Hydrocarbons are said to be hydrophobic since they do not easily dissolve in water.

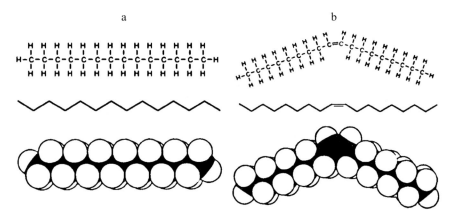

Fig. 2.1. Hydrocarbon chains shown in three different representations. *Top*: all atoms and all bonds. *Middle*: bonds between invisible carbon atoms placed at the vertices. *Bottom*: space-filling models. (**a**) Saturated hydrocarbon chain with fourteen carbon atoms. (**b**) Mono-unsaturated hydrocarbon chain with eighteen carbon atoms. The double bond is here positioned in the middle of the chain

A hydrocarbon chain can be turned into a *fatty acid* by attaching a –COOH (carboxyl) group at the end, as shown in Fig. 2.2a. The carboxyl group is said to be hydrophilic since it can be dissolved in water. Fatty acids, therefore, more easily dissolve in water than pure hydrocarbons. The fatty acids are the fundamental building blocks of all lipids in living matter. Plants and animals use a variety of fatty acids with chain lengths ranging from two to thirty-six. The most common chain lengths fall between fourteen and twenty-two. As we shall see in Sect. 15.1, this is likely to be controlled by the need for cells to have membranes with a certain thickness in order to function properly. Some bacteria have been found to have fatty-acid chains as long as eighty. The simplest way of classifying fatty acids is to write the number of carbon atoms followed by the number of double bonds. For example, myristic acid is denoted 14:0, oleic acid 18:1, and docosahexaenoic acid 22:6. Obviously, a more elaborate notation is needed to specify where double bonds are located along the chain. We shall return to this in Sect. 16.1.

It is most common to find chains with an even number of carbon atoms, although odd ones are found in rare cases. In animals and plants, most of the fatty-acid chains are unsaturated, most frequently with a single double bond (e.g., oleic acid shown in Fig. 2.2b) and in some cases with as many as six double bonds (docosahexaenoic acid, DHA), as shown in Fig. 2.2c. Unsaturated fatty acids with more than one double bond are called *polyunsaturated*. Those with as many as five and six are called *super-unsaturated*. The occurrence of poly- and super-unsaturated fatty acids and how they are synthesized are described in Sect. 16.1.

Short-chain fatty acids can be produced by electrical discharges, e.g., lightening, out of inorganic compounds like carbon dioxide and methane. Intermediate- and long-chain fatty acids are believed only to be produced by biochemical synthesis in living organisms. Therefore, these fatty acids, along with amino acids, are taken as signs of life and are hence looked for in the exploration of extraterrestrial life, e.g., in comets and on Mars.

Fatty acids are rarely found free in the cell, except when they transiently appear in the course of chemical reactions or are transported from cell to cell while attached to certain transporter proteins, so-called lipoproteins. Instead, they are chemically linked to a hydrophobic group, e.g., *glycerol*, which is an alcohol that can be esterified in up to three positions, as illustrated in Fig. 2.2d and e in the case of a di-acylglycerol derived from myristic acid and a tri-acylglycerol derived from stearic acid. This process leads to the formation of a lipid molecule, in this case a *non-polar lipid*. The fatty-acid chains at the different positions can be different and most often they are, typically with a middle one that differs from the other two. Glycerol acts as the *backbone* of the lipid molecule. For comparison, a polar lipid molecule, DMPC, is shown in Fig. 2.2f (cf. Sect. 2.2).

When an ester bond is formed, a water molecule is released. The reverse process, where an ester bond is broken, is referred to as *hydrolysis*

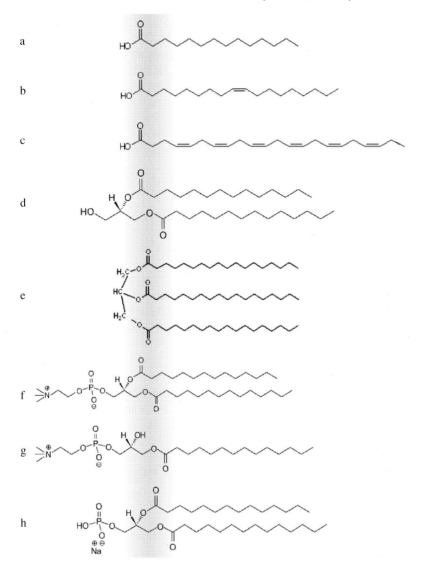

Fig. 2.2a–h. Fats. The polar and aqueous region is shown to the *left* and the hydrophobic region to the *right*. The interfacial region is highlighted in *grey*. (**a**) Fatty acid (myristic acid, 14:0) corresponding to the hydrocarbon chain in Fig. 2.1a. (**b**) Oleic acid (18:1, with one double bond) corresponding to the hydrocarbon chain in Fig. 2.1b. (**c**) Docosahexaenoic acid (DHA) with six double bonds (22:6). (**d**) Di-acylglycerol (DAG) of myristic acid in (a). (**e**) Tri-acylglycerol (triglyceride) of stearic acid. (**f**) Lipid (di-myristoyl phosphatidylcholine, DMPC) made of the di-acylglycerol in (d) and phosphatidylcholine. (**g**) Lysolipid. (**h**) Phosphatidic acid

(i.e., breaking water) or *lipolysis* (i.e., breaking lipids). Certain enzymes can perform lipolysis and we shall return to this in Chap. 18. The result of the lipolysis can be the formation of a so-called *lysolipid* which is a lipid missing one of the fatty-acid chains as shown in Fig. 2.2g. The effect of other enzymes to be described in Sect. 18.1 is to produce di-acylglycerol as shown in Fig. 2.2d or phosphatidic acid as shown Fig. 2.2h.

Di-acylglycerol (DAG) with two fatty acids is a key lipid molecule in certain signaling pathways which we shall describe in Sect. 19.3. Tri-acylglycerols are the typical storage lipid or fat, used for energy production, and saved in certain fat cells (adipocytes) and specialized fat (adipose) tissues.

2.2 The Polar Lipids – Both Head and Tail

Tri-acylglycerols are strongly hydrophobic which means that they cannot be dissolved in water. The affinity for water can be improved by replacing one of the fatty acids with a polar group. The resulting *polar lipid* molecule, called a glycero-phospholipid, then appears as a molecule with a *hydrophobic tail* and a *hydrophilic head* (see also Sect. 3.2). An example is given in Fig. 2.3a. In this case the polar head group is phosphatidylcholine (PC). Other head groups are phosphatidylserine (PS), phosphatidylethanolamine (PE), phosphatidylglycerol (PG), and phosphatidylinositol (PI) as shown in Fig. 2.3.

While PC and PE lipids are neutral (zwitter-ionic), PS, PG, and PI lipids can be electrically charged. This difference has an important consequence for the capacity of the lipids, when incorporated into a lipid membrane, to bind proteins and drugs. The examples of phospholipids shown in Fig. 2.3a–f have two fatty-acid chains that are the same. The lipids in natural membranes usually contain two different chains and most often one of them is unsaturated. In the following, we shall for convenience represent polar lipids by the simple schematic illustration in Fig. 2.3g.

Nature also uses another strategy to construct lipids with head and tails. Instead of using glycerol to bind the fatty acids, *sphingosine*, which is a long-chain amine, is bound to the fatty acid. The simplest version of the resulting sphingolipid is *ceramide*, shown in Fig. 2.4a. Ceramide is an important component of the skin, as we shall discuss in Sect. 19.1. It is also involved in programmed cell death, which is the topic dealt with in Sect. 19.4. Ceramide is a rather hydrophobic molecule. A head group can be attached to ceramide, e.g., PC, leading to *sphingomyelin*, shown in Fig. 2.4b.

The phospholipids, both those based on glycerol and sphingosine, can be installed in the head group with sugar groups of varying degrees of complexity. Examples are shown in Fig. 2.3f and Fig. 2.4c and d, respectively. Such lipids are called *glycolipids*.

Phospholipids can be broken down into their different parts by specific enzymes just like tri-acylglycerols can be hydrolyzed. These enzymes are

2.2 The Polar Lipids – Both Head and Tail

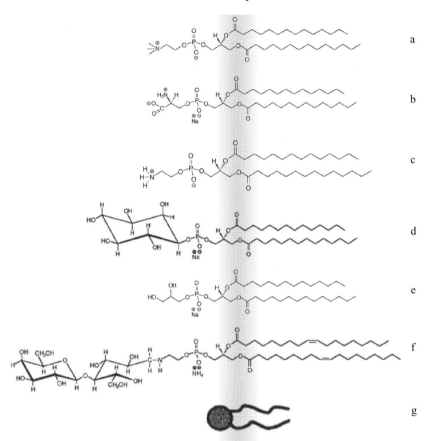

Fig. 2.3a–g. Different polar head groups of glycero-phospholipids. The polar and aqueous region is shown to the *left* and the hydrophobic region to the *right*. The interfacial region is highlighted in *grey*. (**a**) Di-myristoyl phosphatidylcholine (PC). (**b**) Di-myristoyl phosphatidylserine (PS). (**c**) Di-myristoyl phosphatidylethanolamine (PE). (**d**) Di-myristoyl phosphatidylinositol (PI). (**e**) Di-myristoyl phosphatidylglycerol (PG). (**f**) A glycolipid. (**g**) Schematic representation of a polar lipid with a hydrophilic head group and a hydrophobic tail consisting of two hydrocarbon chains

called phospholipases, and their modes of action will be described in Chap. 18. Another type of enzyme, sphingomyelinase, can hydrolyze sphingomyelin.

Whereas the tri-acylglycerols are storage and fuel lipids, phospholipids and sphingolipids are structural and functional lipids. An enormous range of possible lipids can be perceived by varying, e.g., fatty-acid chain length, degree of saturation, polar head group, and type of glycosylation. It is hence not surprising that lipids are the most chemically diverse group of molecules

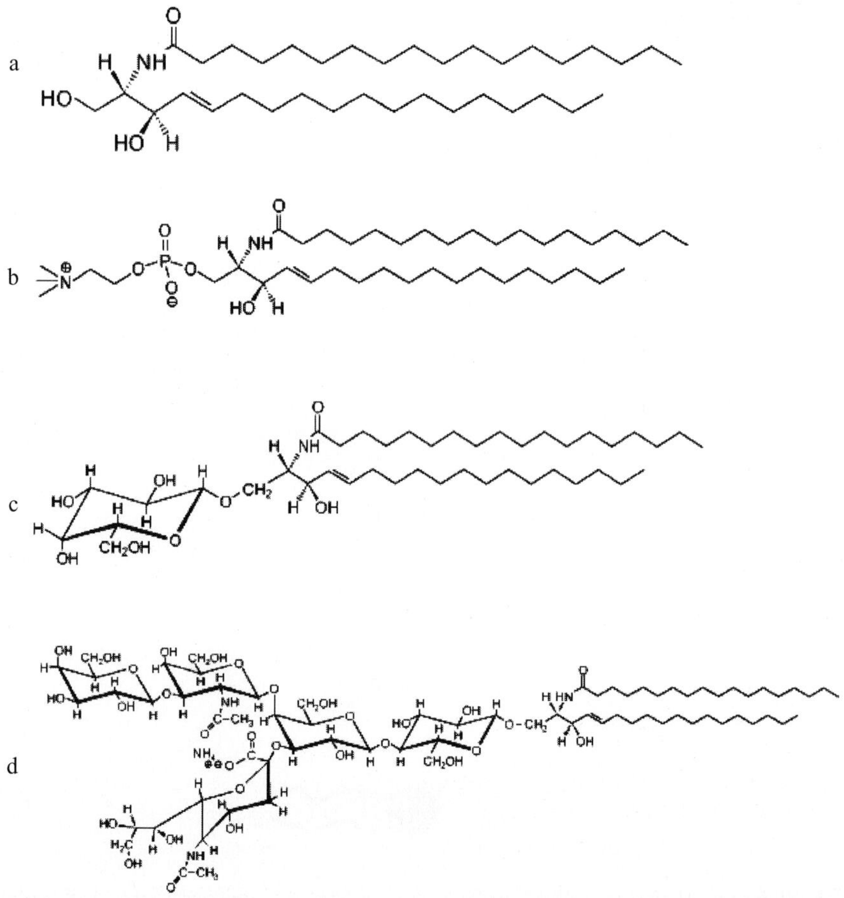

Fig. 2.4a–d. Phospholipids based on sphingosine. The different molecules are not drawn to scale. (**a**) Ceramide. (**b**) Sphingomyelin (SM). (**c**) Cerebroside. (**d**) Ganglioside

in cells. The question naturally arises as to what is the reason and need for this richness and diversity?

2.3 Cholesterol – A Lipid of Its Own

Cholesterol is a lipid that is quite different from the phospholipids and sphingolipids we discussed above. Rather than having a fatty-acid chain as its hydrophobic part, cholesterol has a steroid ring structure and a simple hydroxyl group (–OH) as its polar head. The steroid skeleton has a small hydrocarbon chain at the end. Hence, cholesterol can be characterized as a lipid molecule with a bulky and stiff tail and a small head, as shown in Fig. 2.5a.

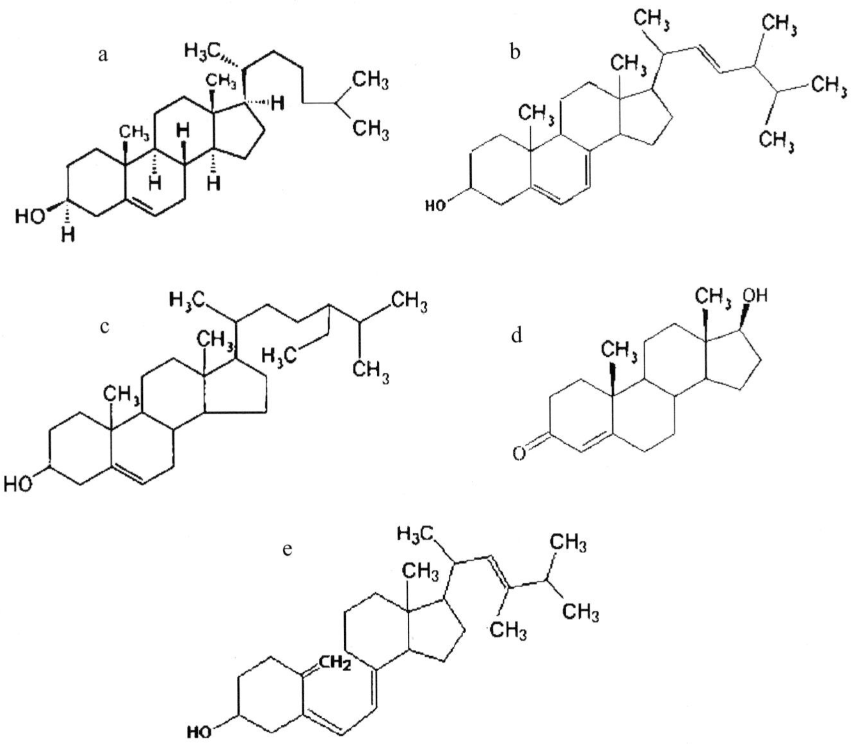

Fig. 2.5a–e. Sterols related to cholesterol. (**a**) Cholesterol. (**b**) Ergosterol. (**c**) Sitosterol. (**d**) Testosterone (male sex hormone). (**e**) Vitamin D

The molecular structure of cholesterol is very similar to that of bile salt, vitamin D, and sex hormones. Cholesterol is one of several members of the sterol family that play similar roles in different types of organisms, e.g., ergosterol in fungi and sitosterol in plants, cf. Fig. 2.5.

2.4 Strange Lipids

Some lipids appear to have rather strange structures that may suggest that they are useful for optimizing the physical properties of membranes that have to work under unusual conditions, for example, at deep sea or in hot springs, as described in Sect. 19.2. These lipids are either very bulky, very long, or based on ether chemistry rather than ester chemistry. The fact that we consider these lipids as strange is likely to reflect that the current fashion of research is biased toward eukaryotic, in particular, mammalian membranes, and that the world of, e.g., the eubacteria and the archaebacteria is much less explored.

In Fig. 2.6 are listed several lipids with unusual structures. Cardiolipin in Fig. 2.6a is basically a dimer lipid that has four fatty-acid chains and is found in the inner mitochondrial membrane, in plant chloroplast membranes, as well as in some bacterial membranes. Lipids based on ether bonding of fatty acids rather than ester bonding are frequently found in archaebacterial membranes. As an example, a di-ether lipid with branched fatty-acid chains is shown in Fig. 2.6b. *Bolalipids* refer to a class of bipolar lipids, i.e., lipids with a polar head in both ends, that can span across a bacterial membrane. Figure 2.6c shows an example of a bolalipid being a tetra-ether lipid, which is a basic component of the membranes of halophilic archaebacteria.

Finally, poly-isoprenoid lipids, as illustrated in Fig. 2.6d, are commonly associated with both prokaryotic and eukaryotic membranes and can act as lipid and sugar carriers.

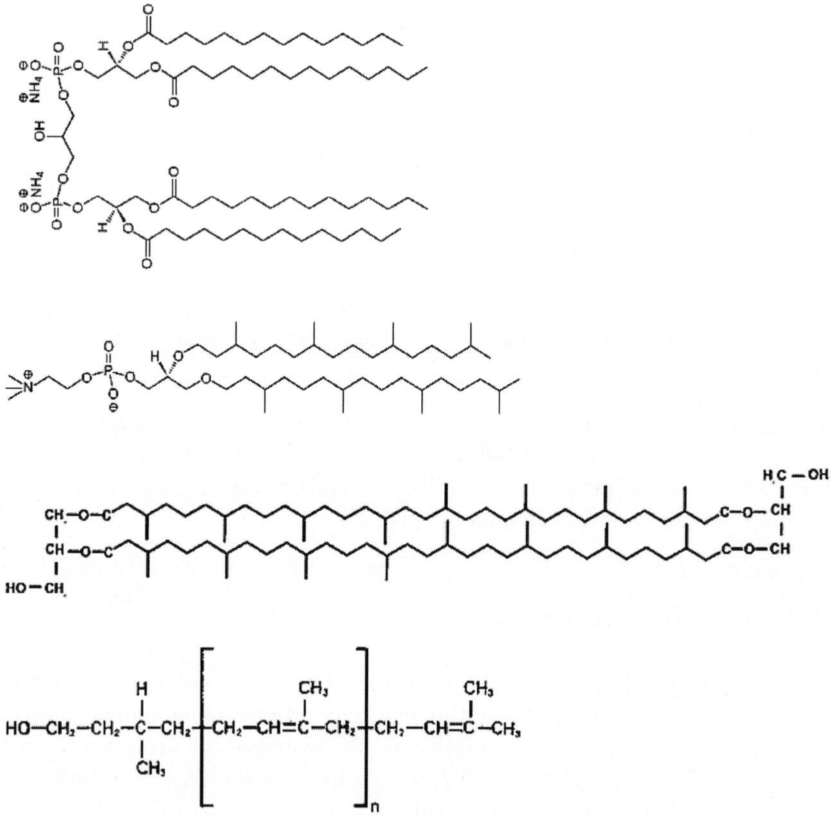

Fig. 2.6a–d. A selection of strange lipids. (**a**) Cardiolipin. (**b**) Di-ether lipid. (**c**) Tetra-ether lipid (bolalipid, or di-biphytanyl-diglycerol-tetraether). (**d**) Poly-isoprenoid lipid

2.5 Lipid Composition of Membranes

As suggested by the description above, an enormous range of different lipids can be constructed. Obviously, nature only exploits some of the possibilities, although the number of different lipid species found in a given kingdom of life, and even within a single cell type, is surprisingly large. Furthermore, a given type of cell or organism can only synthesize a limited range of lipids. For example, human beings only produce a few types of fats and lipids themselves. Most of the fats and lipids in our bodies come from the diet. We shall discuss in Chap. 16 this issue in the context of the fats of the brain and in the visual system.

Without making an attempt to give an overview of the lipid contents in different organisms and cell types, we quote some striking observations for mammalian plasma membranes that shall turn out to be relevant when discussing the physics of lipid membranes. It should be noted that there are characteristic differences between the lipid composition of plasma membranes and that of the various organelles.

Cholesterol is universally present in the plasma membranes of all animals (sitosterol in plants) in amounts ranging between 20–50% of total lipids. In contrast, the organelle membranes contain very little: mitochondrial membranes, less than 5%; Golgi membranes, about 8%; and ER membranes around 10%. In contrast, sterols are universally absent in the membranes of all prokaryotes. These striking numbers can be related to the role played by cholesterol in the evolution of higher organisms, as described in Sect. 14.2.

The amount of charged lipids is about 10% of total lipids in plasma membranes, but there is a substantial variation in the ratio between PS and PI lipids. It is a remarkable observation that nature only uses negatively charged and not positively charged lipids in membranes. It is generally found that the longer the fatty-acid chain, the more double bonds are present. For example, lipids with eighteen carbon atoms have typically one double bond, those with twenty have four, and those with twenty-two have six. PC lipids have typically short chains, whereas SM often have very long chains. PE, PS, and PI lipids typically carry a high degree of unsaturation, whereas PC carry less.

There is a remarkable *lipid asymmetry* in the lipid composition of the two monolayers of the bilayer of the plasma membrane. Whereas SM, PC, cholesterol, and glycolipids are enriched in the outer monolayer, PS, PI, and PE are enriched in the inner layer.

3 Oil and Water

3.1 Water – The Biological Solvent

Water is necessary for life of the form we know. In fact, it is so essential that when NASA goes into space and looks for signs of extraterrestrial life, the search is concentrated on water and features of planetary surfaces that may reflect that water is present or has been present. Moreover, it is mandatory for the evolution of life and maintenance of life processes that water be present in its liquid state. This is why water is called the biological solvent.

No other solvent can substitute for water in supporting life. The reason is found in the peculiar properties of the water molecule and how it interacts with other water molecules in condensed phases like liquid water and ice. Water molecules have a unique capacity for forming *hydrogen bonds* in addition to ordinary dipole-dipole interactions. A hydrogen bond is formed when a hydrogen atom in one water molecule is attracted by a nonbonding pair of electrons from the oxygen atom of a neighboring water molecule, as illustrated in Fig. 3.1. Since each oxygen atom can contribute to two hydrogen bonds, every water molecule can participate in up to four hydrogen bonds with its neighbors. By this mechanism, liquid water is said to form a loose network of hydrogen bonds. It is the peculiar properties of this network that provide the driving force for forming the organized structures and self-assembled materials that are the basis of all life, whether it be the specific molecular structure of a protein or the self-assembly of lipids into a bilayer membrane.

In liquid water, the hydrogen bonds are constantly formed and broken, leading to a very dynamic situation. Due to the many different ways the hydrogen network can be dynamically arranged, the network is strongly stabilized by entropy. This is the reason why water is such a coherent liquid with both a high surface tension and a high boiling point compared with other hydrogen compounds with a similar molar mass, e.g., methane or ammonia.

The hydrogen bonding network is a property that emerges because many water molecules act in concert. The entropy that stabilizes the structure is a nonlocal quantity, i.e., it is not a property of the individual water molecule but emerges from the cooperative nature of a large collection of water molecules.

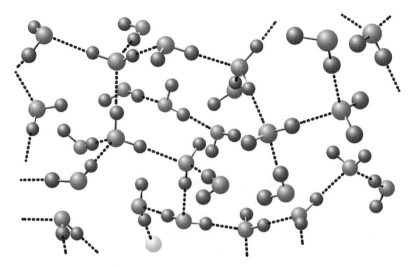

Fig. 3.1. A collection of water molecules. Each water molecule can participate in up to four hydrogen bonds indicated by *dotted* lines

3.2 The Hydrophobic Effect

The stability of the hydrogen bonding network in water makes it difficult to dissolve oil and oil-like compounds in liquid water. Materials that are difficult to dissolve in water are called *hydrophobic* compounds, i.e., compounds that are "water haters." It is well known that oil and water do not mix. This is referred to as the *hydrophobic effect*. Typical oil molecules are simple hydrocarbon chains, as shown in Fig. 3.2. Oil molecules cannot form hydrogen bonds, and liquid oil is therefore only held together by dipole-dipole interactions.

The reason why oil does not mix with water is not so much that the individual parts of the hydrocarbon molecules do not interact favorably with the water molecules via dipole-dipole interactions but that oil is not capable of forming hydrogen bonds. Hence, when an oil molecule is put into water, the hydrogen bonding network in water suffers. As a consequence, the entropy is lowered and the stability of the whole system is decreased. The hydrophobic effect, which will act to drive the oil molecules together in order to diminish the contact with water, is therefore to a large extent of entropic origin.

Compounds that can be dissolved in water are called *hydrophilic*, i.e., "water lovers." Examples are polar or ionic compounds, which due to their charges can form hydrogen bonds.

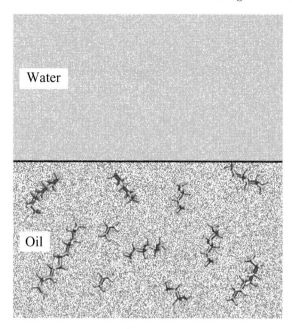

Fig. 3.2. Example of oil molecules (hydrocarbon chains of different length) that in the presence of an oil-water mixture partition into the oil phase due to the hydrophobic effect

3.3 Mediating Oil and Water

It is well known that water and oil can be made to mix if appropriate additives are used. For example, olive oil and vinegar can be mixed to mayonnaise if a stitch of egg yolk or egg white is applied. Similarly, greasy fat in textiles or on our skin can be removed by water if soaps or detergents are brought into use. The ability of these compounds to mediate oil and water can be appreciated if one considers the energetics of the interface that are formed between oil and water as a result of the hydrophobic effect. The interface is characterized by an *interfacial tension* (or surface tension), γ, which is a measure of the free energy that is required to increase the interface between oil and water (cf. Sect. 5.2). Obviously, the larger γ is, the more unfavorable it is to form an interface.

The interface between oil and water can be mediated, and the interfacial tension lowered, if one introduces compounds that are water-like in one end and oil-like in the other end. Molecules with these combined properties are called *amphiphilic* or *amphiphatic*, i.e., they love both oil and water and therefore have mixed feelings about water. The stuff in egg yolk and egg white has such qualities; in fact, egg yolk consists of amphiphilic proteins and lecithin, which is a mixture of different lipid molecules. Similarly, soaps and detergents are also amphiphilic, typically salts of fatty acids.

Due to their mixed feelings about water, amphiphilic molecules tend to accumulate in the oil-water interface, as shown in Fig. 3.3. This leads to a lowering of the interfacial tension, which, in turn, facilitates the mixing of oil and water. The resulting mixture is called an emulsion or dispersion. Amphiphilic molecules are also called interfacially active compounds, emulgators, or surfactants. They are of extreme technological importance not only in detergency, but also for processing foods and for making cosmetics, paints, and surface modifiers. Examples of interfaces in oil/water emulsions are shown in Fig. 3.3c–g. The organization of amphiphilic molecules at the interfaces of the emulsion is a consequence of many molecules acting in unison.

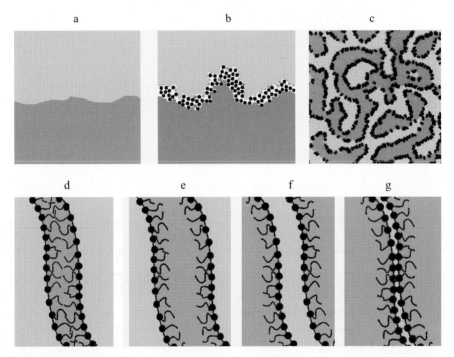

Fig. 3.3. A single interface between oil and water in the absence (**a**) and the presence (**b**) of interfacially active molecules. (**c**) A fully developed oil-water emulsion with a collection of interfaces. Oil/water interfaces of an oil-in-water mixture (**d–e**) and a water-in-oil mixture (**f–g**)

Nature has long ago discovered that amphiphilic molecules in the form of lipids are indispensable for "emulgating" living matter. Lipids are nature's own surfactants. It is the amphiphilic character of lipids that give them a unique position among the molecules of life.

3.4 Self-Assembly and the Lipid Aggregate Family

When mixing lipids with water, the hydrophobic effect acts to make sure that the oily chains of the lipid molecules are screened as much as possible from water. This leads to a whole family of supramolecular aggregates that are formed spontaneously by self-assembly. The self-assembly process is a many-molecule effect, and it requires that many lipid molecules act together. The family of lipid aggregates is illustrated in Fig. 3.4. In all these lipid aggregates, the polar head of the lipids is hydrated by water and the fatty-acid chains are tugged away from the water.

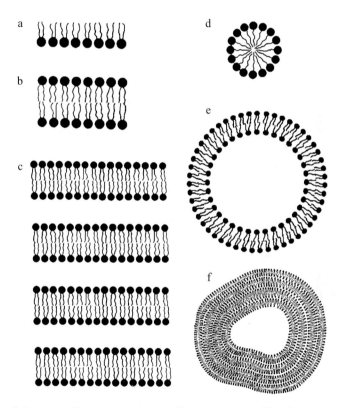

Fig. 3.4. Schematic illustration of the self-organization of lipids into supramolecular aggregates in association with water. (**a**) Lipid monolayer. (**b**) Lipid bilayer. (**c**) Multi-lamellar lipid bilayers in a stack. (**d**) Micelle. (**e**) Vesicle or liposome (closed lipid bilayer). (**f**) Multi-lamellar vesicle or liposome

The simplest and most ideal lipid aggregate form is the lipid *monolayer* in Fig. 3.4a. The lipid monolayer is a monomolecular film of lipids formed on the interface between water and air (or another hydrophobic substance, such as oil). We shall describe lipid monolayers in more detail in Chap. 10.

The lipid bilayer in Fig. 3.4b can be considered as two monolayers back to back like in a skinny oil-in-water mixture (cf. Fig. 3.3d). Several lipid bilayers often organize among themselves to form multi-lamellar structures, as shown in Fig. 3.4c. The forces that stabilize a stack of bilayers are of a subtle origin and will be dealt with in Sect. 5.3. Obviously, open ends cannot be tolerated, and the lipid bilayers have to close onto themselves and form closed objects, as shown in Fig. 3.4e and f. Such structures are called respectively uni-lamellar and multi-lamellar lipid *vesicles* or *liposomes*.

Images of vesicles and liposomes obtained by microscopy techniques are shown in Fig. 3.5. A uni-lamellar liposome constitutes the simplest possible model of a cell membrane. It should be noted that lipids extracted from biological membranes when mixed with water will self-assemble and can form lamellar lipid aggregates, as in Fig. 3.5, although they quite often form the non-lamellar structures described in Sect. 4.3.

The lipid aggregates in Fig. 3.5 are all characterized by being of planar or lamellar symmetry. This requires that the lipid molecules have a shape that is approximately cylindrical in order to fit in. If the shape is more conical, other aggregate symmetries may arise of curved form as described in Chap. 4. Nonpolar lipids like triglyceride oil do not form aggregates in water, whereas all polar lipids, except cholesterol, form aggregates in water.

The most important lesson from the observation of lipid self-assembly is that lipid aggregates, e.g., lipid bilayer membranes and hence biological membranes, owe their existence to water as the biological solvent. The aggregate and the solvent are inextricably connected. Lipid bilayers do not exist on their own in the absence of water. Moreover, the fact that lipid aggregates are formed and stabilized by self-assembly processes implies that they possess self-healing properties. If they are subject to damage, e.g., hole- or pore-formation in lipid bilayers, the damage is often repaired automatically by filling in holes and by annealing various defects.

It is instructive at this point to compare the self-organization and formation of lipid structures in water with another important self-assembly process of immense importance in biology, the process of *protein folding*, as illustrated in Fig. 3.6a–b. Most proteins are composed of both hydrophobic and hydrophilic amino acids. When exposed to water, proteins will undergo a folding process, which leads to a molecular structure that is a compromise between minimizing the exposure of the hydrophobic amino acid residues to water on the side and maximizing the interactions between the various amino acids in the sequence. These interactions involve electrostatic forces, hydrogen bonds, as well as sulphur bridges. The resulting structure is a delicate balance between these forces. Hence, the structure of the protein, and therefore also its function, is very sensitive to shifting this balance by changes in external conditions, e.g., temperature or pH. Such changes can induce a complete or partial unfolding of the structure, also termed denaturation. Many proteins are water soluble and exert their function in water. The folding process of

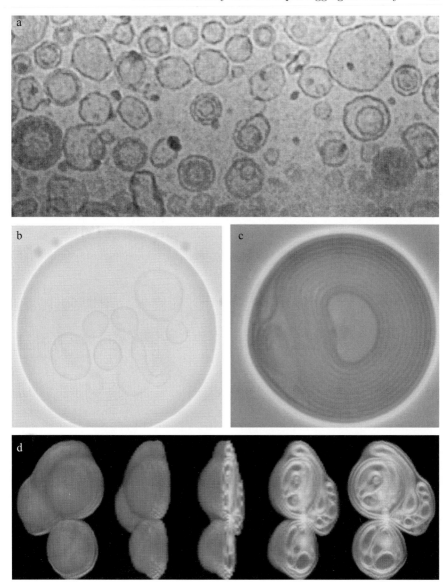

Fig. 3.5a–d. Uni-lamellar and multi-lamellar liposomes as obtained by microscopy techniques. (**a**) Small uni-lamellar liposomes of approximately 100 nm diameter. The picture is obtained by cryo-electron microscopy. Smaller vesicles are seen to be trapped inside some of the liposomes. (**b**) A large uni-lamellar liposome with a diameter around 70 μm. Smaller uni-lamellar liposomes are trapped inside. (**c**) A large multi-lamellar liposome with an outer diameter of 40 μm. (**d**) Cross section through an agglomerate of multi-lamellar vesicles shown from different angles

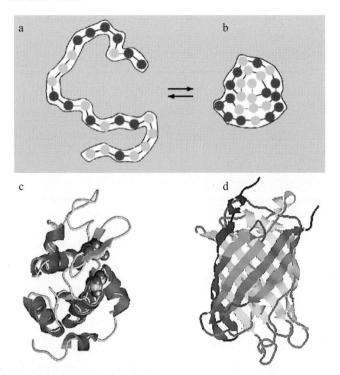

Fig. 3.6. (**a,b**) Schematic illustration of the folding process in water of a protein molecule, represented as a pear necklace, from the unfolded state (**a**) to the folded state (**b**). Hydrophilic and hydrophobic amino-acid residues are highlighted differently. (**c,d**) Two examples of protein structures. (**c**) Lysozyme as an example of a water-soluble protein with predominantly α-helix structure. (**d**) Porin as an example of a membrane-spanning protein with predominantly β-sheet structure

trans-membrane proteins is subject to the extra complication that the folded protein has to come to terms with an environment that is both hydrophilic and hydrophobic. We shall return to this in Sect. 13.3.

3.5 Plucking Lipids

The stability of liquid water provided by the dynamic hydrogen bonding network has as a consequence that liquid water is more dense than solid water where the hydrogen bonding network is forced to be more fixed. This explains the well-known observation that ice flows on the top of liquid water. In this respect, water is very different from most other substances where the solid state usually is more dense than the liquid state. This highlights the special role of water as the only possible biological solvent. If ice would sink to the bottom of lakes and oceans, life as we know it could not have evolved and survived under the climate conditions of planet Earth.

When dissolving amphiphilic and hydrophobic material like lipids and proteins in water, the water molecules closest to the hydrophobic material will be unable to engage in all four possible hydrogen bonds and will suffer from a change in the hydrogen bonding dynamics. This implies that the part of the water that can "feel" the hydrophobic material will be more structured and less dense than bulk liquid water. In other words, the water is depleted from the hydrophobic surface, which becomes effectively de-wetted. The density of this structured water is similar to that of amorphous ice, which is about 90% of bulk liquid water. This reduction in water density has, in fact, been observed in an experiment that can measure the density of water close to a layer of hydrocarbons. The results have been confirmed by computer-simulation calculations, as shown in Fig. 3.7. Figure 3.7 also leads to an estimate of the range over which water is perturbed by a hydrophobic surface. This range is about 1–1.5 nm.

The hydrophobic effect is accompanied by subtle changes in the ordering of the water dipole moments. In contrast to common belief, recent work

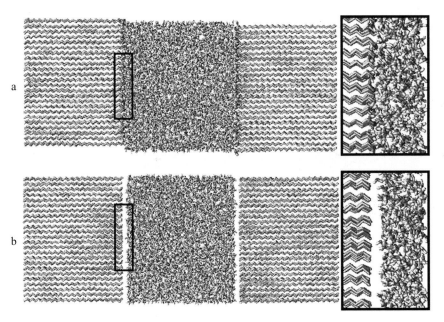

Fig. 3.7. A slab of water between two solid surfaces made of (**a**) a long-chain alcohol and (**b**) a long-chain hydrocarbon. The density of water near the hydrophobic surface in (**b**) is seen to be smaller. To the *left* are shown enlarged versions of the interfacial region illustrating how the hydrophilic alcohol surface is wetted by water and the hydrophobic hydrocarbon surface is de-wetted. The de-wetting is a direct manifestation of the hydrophobic effect

has shown that the water dipoles are more ordered near hydrophilic surfaces than hydrophobic surfaces. The higher degree of water ordering at hydrophilic surfaces implies a slowing down of the water diffusion along the interface.

The hydrophobic effect is usually quantified by assigning a so-called transfer free energy to the process of transferring hydrophobic molecules from a hydrophobic, oily phase into water. The Canadian biophysicist Evan Evans has succeeded in monitoring this transfer process on the level of a single molecule by "plucking" a single lipid molecule from a lipid bilayer membrane. This unique type of experiment is illustrated in Fig. 3.8. A single lipid molecule is targeted by binding of a receptor molecule (avidin) to a ligand (biotin) that is chemically bound to the head group of the lipid molecule. The receptor molecule, in turn, is linked to a micrometer-sized glass bead. The glass bead, in turn, is attached to the surface of a soft body, such as a swollen red blood cell or a liposome. This soft body acts like a spring whose spring constant can be varied by changing the tension of the body. This is done by pressurization, using a micro-pipette as shown in the figure. The resulting mechanic transducer, which is called a bioprobe force spectrometer, can be used to measure the force exerted on the soft body, e.g., when it becomes distorted during extraction of the lipid molecule from the target membrane, as illustrated in Fig. 3.8b. The force it takes to extract the lipid molecule depends on the rate by which the pulling is performed. Except when extracted extremely rapidly, the anchoring strength of a lipid molecule is very weak, typically 2–4 pN, thus providing a quantitative measure of the hydrophobic effect.

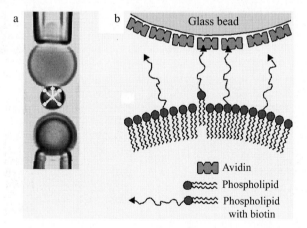

Fig. 3.8. Extracting a single lipid molecule from a lipid bilayer using a bioprobe single-molecule force spectrometer developed by Evan Evans. (**a**) The spectrometer involves a micro-pipette, a soft body, e.g., a red blood cell or a liposome, and a glass bead (here shown with a cross). (**b**) The target lipid molecules in the membrane are associated with biotin moieties that can be chemically bound to the avidin receptor molecules linked to the glass bead

4 Lipids Speak the Language of Curvature

4.1 How Large Is a Lipid Molecule?

The dimensions of a lipid molecule is determined by several factors. Firstly, there are obvious geometric factors like the size of the polar head, the length of the fatty acid tail, and the degree of unsaturation of the fatty-acid chains. Figure 4.1 shows examples where the molecules are inscribed by cylinders. Obviously, the longer the fatty acid tail is, the longer is the hydrophobic part of the molecule. The chains in this figure are stretched out as much as they can be. In the case of one or more double bonds, the end-to-end length of a chain will be shorter than for chains with fewer double bonds and the same number of carbon atoms. Double bonds will make the chain depart from the linear arrangement, as illustrated in Fig. 4.1c, and the approximation by a cylinder will be less good. For a given number of double bonds, the length of the hydrophobic part of a lipid molecule (and consequently the thickness of the lipid bilayer it may form, cf. Sect. 8.3) is linearly proportional to the number of carbon atoms in the chains. To illustrate this fact and for later reference, Fig. 4.2 shows a homologous family of di-acyl PC with two identical saturated chains.

The actual conformation of the molecule will influence its effective size. A *conformation* refers to the actual spatial arrangements of the atoms of the molecule. In Figs. 4.1 and 4.2 are drawn the conformations that have the lowest conformational energy, i.e., very ordered conformations in which all the C–C–C bonds occur in a zigzag arrangement (all-*trans*). However, temperature effect will lead to rotations, so-called excitations, around the C–C bonds and consequently to more disordered conformations. In this sense, lipids are qualitatively different from the other energy-producing molecules of the cell, the carbohydrates, which frequently are composed of stiff ring structures that allow for limited flexibility.

An example of a series of excited conformations of lipid molecules in a bilayer is shown in Fig. 4.3. It is clear from this figure that the long flexible chains of the lipids imply that the effective size and shape of the lipid molecules are very dependent on temperature. This property is of immense importance for the use of lipids in biological membranes. It is the source of the softness of lipid membranes, which is needed for their function.

44   4 Lipids Speak the Language of Curvature

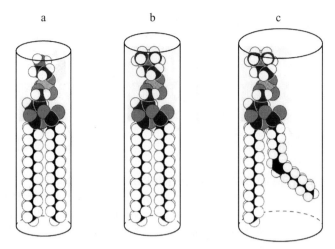

Fig. 4.1a–c. Schematic illustration of the dimensions of lipid molecules. (**a**) Di-stearoyl phosphatidylethanolamine (DSPE). (**b**) Di-stearoyl phosphatidylcholine (DSPC). (**c**) Stearoyl-oleoyl phosphatidylcholine (SOPC)

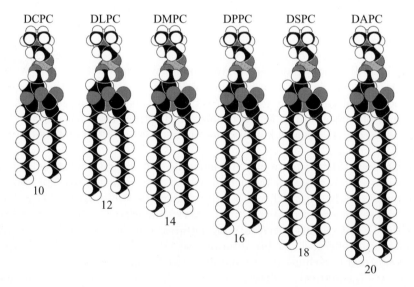

Fig. 4.2. The homologous family of di-acyl PC lipids with two identical saturated chains. The figure also serves to define the acronyms traditionally used for these lipids. The numbers at the bottom denote the number of carbon atoms in the fatty-acid chains

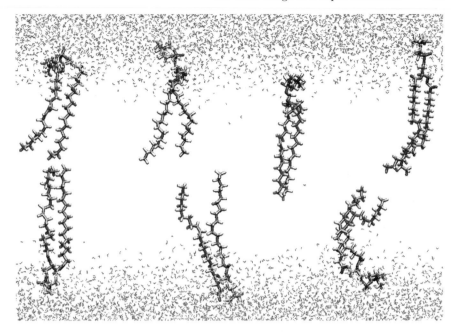

Fig. 4.3. Some conformations of SOPC molecules in a lipid bilayer structure as obtained from computer-simulation calculations. For clarity, only some of the lipids in the two monolayer leaflets are shown. The water molecules are shown as small angles

Only lipids with a limited degree of disorder will fit into a bilayer structure. In general, the average molecular shape has to be close to that of a cylindrical rod, as we shall see in Sect. 4.2 below. In a lipid-bilayer membrane in the physiological state, the typical cross-sectional area of this cylinder is about $0.63\,\mathrm{nm}^2$ and its average length from 1.0–1.5 nm, depending on the chemical nature of the fatty-acid chains, in particular, the number of carbon atoms and the degree of saturation. The average length of the chains determines the hydrophobic thickness of the bilayer membranes it can form, as described in Sect. 8.3.

In this context, it should be remarked that there are two ways of forming double C–C bonds: *cis*-double bonds and *trans*-double bond. Nature usually makes *cis* double bonds in fatty acids. The *trans*-double bond leads to a less jagged chain, which has a significant ordering effect on the membrane lipids. This difference is part of the reason why *trans*-fatty acids in foods are less healthy than *cis*-fatty acids.

4.2 Lipid Molecules Have Shape

It may already have been noticed from Fig. 4.3 that temperature has an effect not only on the size of a lipid molecule but possibly also on its shape. The effective molecular shape is important for the ability of a lipid to form and participate in a bilayer structure. It is a matter of fitting. A word of caution is in order at this point. The use of the term shape can be misleading if taken too literally. A lipid molecule, when incorporated into a lipid aggregate like a bilayer, does not occupy a well-defined volume of a well-defined shape. At best the effective shape of a lipid molecule describes how its average cross-sectional area depends on how deeply it is buried in the lipid aggregate. Therefore, the effective shape is a property that is influenced by the geometrical constraints imposed by the aggregate. This will become more clear when in Sect. 8.2 we describe the various forces that act in a lipid bilayer. With this caveat we shall take the liberty to assign an effective shape to lipid molecules.

It has in recent years become increasingly clear that lipid shape is important for functioning. We shall demonstrate this by a specific example in Sect. 4.4 below and return to the mechanism of coupling curvature to protein function in Sect. 15.2.

The effective shape of a lipid molecule is determined by the compatibility between the size of the head group and the size of the hydrophobic tail. Full compatibility implies an effective cylindrical form. The effective shape of a lipid molecule, as a measure of its ability to fit into a particular lipid aggregate, is conveniently described by a packing parameter

$$P = \frac{v}{a \cdot l}, \tag{4.1}$$

where v, a, and l are defined in Fig. 4.4. Since the volume of a cylinder-shaped molecule is $a \cdot l$, a deviation of P from unity suggests that non-lamellar aggregates can be expected. $P > 1$ corresponds to a shift from a cylindrical shape toward an inverted cone, whereas $P < 1$ corresponds to a shift toward a normal cone.

There are various ways of changing the effective shape of a lipid molecule by varying the relative sizes of the head and tail. A small head and a bulky tail and a large head and a skinny tail will produce conical shapes of different sense, as illustrated in Fig. 4.5. As we shall see later, this variability, which is peculiar for lipids, is used in a wide range of membrane processes.

4.3 Lipid Structures with Curvature

The effective shape of lipid molecules determines their ability to form a stable bilayer. The more non-cylindrical their shapes are, the less stable a bilayer they will form. This is illustrated in Fig. 4.6, where the two monolayers possess separately an intrinsic tendency to elastically relax toward a state of

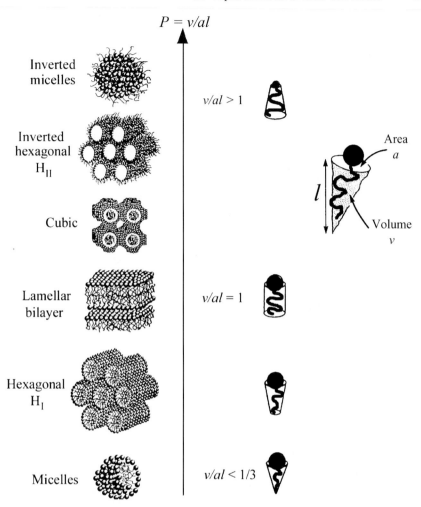

Fig. 4.4. Schematic illustration of lamellar and non-lamellar lipid aggregates formed in water. The different structures have different curvature and are arranged in accordance with the value of packing parameter $P = v/al$

finite curvature. The monolayers display a so-called *spontaneous curvature*. We shall return with a fuller description of spontaneous curvature in Sect. 6.2 in which we consider membranes as mathematical surfaces. When a bilayer is made of monolayers with nonzero spontaneous curvature, it becomes subject to a built-in frustration termed a *curvature stress field*. If the spontaneous curvatures of the two monolayers are different, the bilayer becomes asymmetric and itself assumes a nonzero spontaneous curvature. This is similar to a normal plastic tube, which, when cut open, will maintain its curved shape, whereas a piece of paper wrapped unto itself will not.

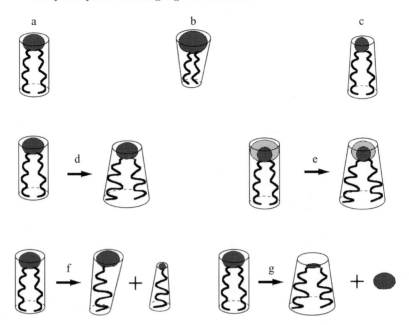

Fig. 4.5a–g. Effective shapes of lipid molecules. (**a**) Cylindrical: similar sizes of head and tail. (**b**) Cone: big head and skinny tail. (**c**) Inverted cone: small head and bulky tail (e.g., with unsaturated fatty-acid chains). (**d**) Going conical by increasing temperature. (**e**) Going conical by changing the effective size of the head group, e.g., by changing the degree of hydration or by changing the effective charge of an ionic head group. (**f**) Going conical by chopping off one fatty-acid chain, e.g., by the action of phospholipase A_2, which forms a lysolipid molecule and a free fatty acid. (**g**) Going conical by chopping off the polar head group, e.g., by the action of phospholipase C

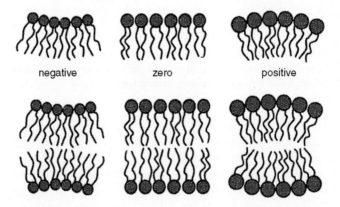

Fig. 4.6. Illustration of the destabilization of a lipid bilayer composed of lipids with conical shapes that promote a tendency for the two monolayers to curve. Bilayers made of monolayers with a nonzero curvature have a built-in curvature stress

If, however, the cohesion of the bilayer cannot sustain the curvature stress, the stress will force non-lamellar structures to form, as shown in Fig. 4.4. Less regular structures known as emulsions and sponge structures formed by lipids at curved interfaces in water will be described in Sect. 5.2. In all these curved structures, the lipids speak the language of curvature. The variety of lipid structures of different morphology is referred to as *lipid polymorphism*.

A particularly interesting structure is the *inverted hexagonal structure* in Fig. 4.4, which in the following will be called the H_{II} structure. This structure is characterized by long cylindrical rods of lipids arranged as water-filled tubes. The diameter of the tubes can be varied by changing the type of lipid and by varying environmental conditions such as temperature, degree of hydration, as well pH. Despite their exotic appearance, both the inverted hexagonal structure, as well as the cubic structure described below, turn out to be of significance for the functioning of biological membranes, as described specifically in Sects. 4.3 and 4.4 and in several other places throughout the book. The other hexagonal structure, H_I, in Fig. 4.4 is not very common, except for very polar lipids and for lipids with very big head groups like lysolipids.

Cubic structures are much more complex than lamellar and hexagonal structures, and there are several types of them. They are bicontinuous in the sense that the water is divided into disconnected regions, one on each side of the lipid bilayer, which is curved everywhere. They are in a mathematical sense so-called minimal surfaces, i.e., surfaces that everywhere are in the form of a saddle with zero mean curvature (cf. Sect. 6.1). Even if cubic structures are subtle and appear exotic, they are related to biological function. It is an interesting observation that cubic phases can dissolve amphiphilic proteins, which can adapt locally to the curvature and, in certain cases, form micro-crystallites. This can be exploited for elucidation of membrane protein structure, as described in Sect. 15.2.

The propensity of the lipids for forming and stabilizing the H_{II} structure is increased by shifting the balance between l and a in 4.1 (cf. Fig. 4.5) – for example, by decreasing the effective size of the head group (by dehydrating the system or screening the charge of the polar head by adding ions), increasing the temperature, increasing the degree of unsaturation of the fatty-acid chains, increasing the length of the fatty-acid chains, or decreasing the hydrostatic pressure. The reason for the last effect, which is important for deep-sea bacteria, is that the volume of lipids in water decreases slightly upon the application of pressure, corresponding to a stretching-out of the fatty-acid chains (cf. Sect. 19.2).

Since the self-assembly process of lipid molecules into aggregates of different morphology implies a subtle competition between forces of different origin and since many of the forces are of a colloidal and entropic nature, the relative stability of the resulting structures is intimately dependent on temperature, composition, and environmental conditions. In particular, increasing temperature can drive a lamellar structure into an inverted hexagonal or cubic structure. Furthermore, incorporation of various hydrophobic and amphiphilic solutes such as hydrocarbons, alcohols, detergents, as well as a variety of drugs, can shift the equilibrium from one structure to another in the series shown in Fig. 4.4. For example, mono-acylglycerols, di-acylglycerols, tri-acylglycerols, alkanes, and fatty acids promote the H_{II} structure, whereas detergents, lyso-phosphatidylcholine, digalactosyl diglyceride, certain antiviral peptides, as well as detergents inhibit the formation of the H_{II} structure.

The shape of the cholesterol molecule in relation to membrane curvature deserves a special remark. Compared to the small head group, which is just a –OH group, the steroid ring structure shown in Fig. 2.5a, although hydrophobically rather smooth, is bulky, thus providing cholesterol with an inverted conical shape. Cholesterol, therefore, displays a propensity for promoting H_{II} structures.

It should be remarked that the curvature of large lipid bilayer objects like the vesicles and liposomes shown in Fig. 3.5, is not caused by intrinsic, curvature stress, as described in the present section. The curvature in these cases is simply caused by the boundary conditions. Due to the hydrophobic effect, the bilayer would have to close onto itself, just like biological membranes have. Apart from very small vesicles, which have a large built-in tension, the curvature of liposomes is typically in the range of micrometers, which is much larger than the radii of curvature involved in hexagonal and cubic structures. For these structures, the radii of curvature are of the same order as the size of the lipid molecules.

It is a remarkable observation that lipid structures formed in water by the lipid extract from real biological membranes that are lamellar, often turn out to form non-lamellar phases despite the fact that the composition of these lipid extracts varies tremendously from organism to organism and among the different cell types for the same organism. In addition, more than half of the lipids naturally present in biological membranes, when studied as individual pure lipids, do not form bilayer phases, but rather cubic or inverted hexagonal structures. This may partly be related to the fact that non-lamellar-forming PE-lipids are abundant in both prokaryotic and eukaryotic cells. For example, 70% of the phospholipids in *Escherichia coli* are PE-lipids. Another striking observation is that the lipid bilayer of natural membranes in many cases is found to be close to a transition from a lamellar structure to a non-lamellar structure. This transition, which is a phase transition, as described in Sect. 9.2, can be triggered globally in bilayers made of lipid extracts from the real membranes by using the principles of shape changes illustrated in Fig. 4.5.

It is not desirable for functional biological membranes to deviate globally from a lamellar symmetry, possibly with the exception of the very curved and convoluted membranes of Golgi and the ER, cf. Fig. 1.5. However, the presence of non-lamellar structures as virtual states leads to a curvature stress field in the membrane. The stress field can in fact be changed enzymatically by specific enzymes that change the H_{II} propensity of the lipids while they reside in the bilayer, e.g., by enzymatically cleaving off the polar head or removing one of the fatty-acid chains. The resulting stress may be released locally, e.g., by changes in the local molecular composition, by binding a protein or hormone, or by budding of the membrane as an initiation of a fusion process. We shall in Sect. 15.2 return to these phenomena in connection with the way lipids can control membrane function.

4.4 Microorganisms' Sense of Curvature

An interesting series of studies have been performed on a simple unicellular organism, *Acholeplasma laidlawii*, which show that the lipid composition of this organism is regulated to preserve spontaneous curvature under diverse living conditions. The results suggest that the propensity for forming an H_{II} structure may be a signal for cell growth.

Acholeplasma laidlawii is a so-called mycoplasma that is even simpler than bacteria. It is deficient in synthesizing fatty acids and therefore has to do with the fatty acids it feeds on. The fatty acid composition of its membrane will therefore reflect which fatty acids the mycoplasma selects from its food. However, *Acholeplasma laidlawii* contains specific enzymes that are able to change the polar head group of the lipids. In particular, it can choose to vary the size of the head group by using different sugar groups. It turns out that the organism regulates, given a specific diet of fatty acids, the ratio between glycolipids with different head group sizes in order to compensate for fatty acids that do not pack well into the membrane due to their shape.

In a series of experiments performed by the Swedish chemists Åke Wieslander, Göran Lindblom and their collaborators, *Acholeplasma laidlawii* was fed mixtures of saturated (palmitic) and mono-unsaturated (oleic) acids in different compositions. Being the tail of a phospholipid, oleic acid is expected to have a larger propensity for forming an H_{II} structure as compared to palmitic acid. These fatty acids were found to be effectively incorporated into the lipids produced by the organism and constituted more than 90% of the fatty acid content of the plasma membrane after adaptation. The total lipid content was then extracted, and when mixed with water it was found to form an H_{II} structure. The spontaneous curvature, C_0, of this structure was measured by X-ray analysis. The result showed that C_0 was nearly constant for a wide range of different compositions of the food. In contrast, the corresponding C_0-values of the pure lipids, which are believed to dominate the spontaneous curvature, were found capable of varying within a much larger region. Moreover, it was found that the ratio of large head groups to small head groups adopted by *Acholeplasma laidlawii* varied almost proportionally to the ratio of oleic acid to palmitic acid. Hence, large contents of bulky oleic chains were compensated by similarly large amounts of large head groups, and vice versa. Consequently, the particular value of the spontaneous curvature must in some way be optional for the functioning and growth conditions of the cell.

Hence, it appears that *Acholeplasma laidlawii* is an organism that is able to homeostatically regulate the lipid composition of its membrane in order to maintain a constant spontaneous curvature and hence a constant propensity for forming non-lamellar structures. This is achieved simply by playing around with the compatibility between the head group size and tail size of the lipids in order to obtain the right molecular shape for packing effectively into the lipid bilayer membrane.

Other microorganisms, including *Escherichia coli*, have been proposed to regulate their membrane properties by a similar principle based on lipid molecular shape and optimal packing. It is at present unknown by which mechanism the lipid synthesis is regulated by the curvature stress field of the membrane and which membrane-bound proteins are involved. We shall in Sect. 15.2 discuss possible physical mechanisms of coupling membrane curvature to the functioning of proteins.

5 A Matter of Softness

5.1 Soft Matter

Most biological matter, like membranes, are soft materials. Soft materials have a number of unusual properties that are very different from those of traditional hard materials such as metals, ceramics, semi-conductors, and composites. Lipid membranes are soft because they are basically structured liquids made of molecules with substantial conformational complexity. At the same time, they have tremendous durability and toughness over ordinary liquids due to the fact that they owe their existence to the self-assembly principles described in Sect. 3.4. As we shall see throughout the remainder of this book, the softness is a requirement of the various modes of function that membranes engage in. Technological applications of soft-matter systems made of lipids will be described in Chap. 20.

Soft materials refer to a vast and ubiquitous class of structured and complex systems that include polymers, supramolecular aggregates, emulsions, colloids, liquid crystals, as well as membranes. Examples from daily life are syrup, ketchup, glue, paint, toothpaste, egg white, and silly putty. All these systems exist in a condensed phase, but none of them can be described unambiguously as a liquid or a solid. As opposed to conventional solid materials, the physical properties of soft materials are largely determined by soft and fluctuating interfaces, the physics of which is dominated by entropy. Softness implies high deformability, but not necessarily high bulk compressibility. Furthermore, soft matter is usually anisotropic and constructed in a hierarchical manner with structure occurring on several different length scales that are connected in subtle ways.

5.2 Soft Interfaces

Figure 5.1 shows a gallery of examples of systems with soft interfaces of increasing complexity. All of these systems are structured liquids, and they are basically a collection of soft fluid interfaces. The interfaces are fluid in the sense that there is no fixed relationship between nearest-neighbor molecules within the interface. They exert no resistance to shear forces. A particularly peculiar structure is the sponge phase, which can be considered a disordered

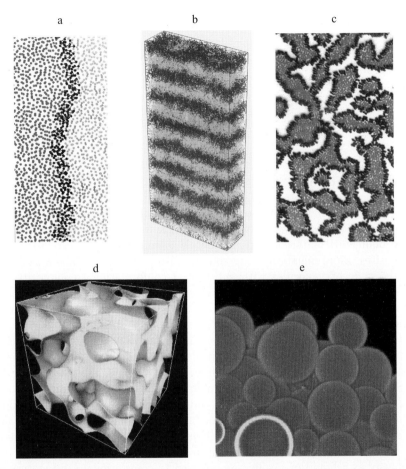

Fig. 5.1a–e. Examples of systems with fluid soft interfaces. (**a**) Liquid-liquid interface enriched in interface-active molecules (e.g., amphiphiles like soaps, detergents, or lipids). (**b**) Di-block co-polymers in a lamellar phase. A di-block co-polymer consists of two incompatible polymers that are chemically linked together. (**c**) Microemulsion, which is a complex collection of convoluted and fluctuating interfaces covered by interface-active molecules like lipids. (**d**) Sponge phase, which is a disordered variant of the bicontinuous cubic lipid phase in Fig. 4.4. (**e**) A collection of lipid vesicles

variant of the cubic phase. The sponge phase is also bicontinuous and can be compared with a complex arrangement of tubes. This has led to its nickname, "the plumber's nightmare." It consists of curved lipid bilayers, but it is not a liquid crystal. We shall have a closer look at this phase in Sect. 6.1.

For comparison, Fig. 5.2 shows a related set of soft matter systems. These systems are also characterized by interfaces. However, in contrast to the interfaces in Fig. 5.1, the interfaces in Fig. 5.2 share some of their internal

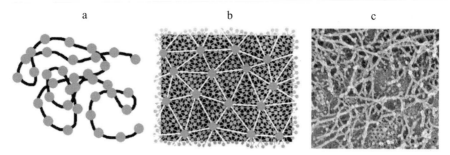

Fig. 5.2a–c. Examples of soft matter systems with tethered interfaces. (**a**) A polymer chain characterized by a tethered string of beads with fixed connections. (**b**) A tethered two-dimensional membrane resembling a cytoskeleton with fixed connectivity attached to a fluid lipid bilayer with dynamically changing connectivity. (**c**) Electron microscopy image of the spectrin network, which is part of the cytoskeleton of the red blood cell. The size of the image is about 500 nm × 500 nm

properties with solids. They are all tethered, implying that they have a fixed relationship between neighboring molecules and therefore display shear resistance. The polymer chain in Fig. 5.2a is a one-dimensional tethered string in which neighboring monomers are bound by chemical bonds. Although flexible on long length scales, the polymer displays some rigidity at a shorter scale, referred to as the *persistence length*. The tethered network in Fig. 5.2b is a two-dimensional generalization of a polymer. Each node in this network is tethered to a certain number of its neighbors. Also, this sheet is flexible but provides a certain shear resistance. A tethered interface is like a sheet of vulcanized rubber. The mechanical, conformational, and statistical properties of tethered interfaces are very different from those of fluid interfaces. In Fig. 5.2c is shown a biological realization of such a tethered network, the spectrin skeleton of a red blood cell. Spectrin is a protein that can cross-link with other spectrin molecules into a two-dimensional network. It serves here to provide the blood cell with its characteristic shape. We shall come back to this in Sect. 6.4.

The properties of fluid interfaces and surfaces are most often controlled by the *interfacial tension*

$$\gamma = \left(\frac{\partial G^{\mathrm{s}}}{\partial A}\right)_V , \qquad (5.1)$$

where G^{s} is the Gibbs excess free energy, V is the volume, and A is the area of the interface. The interfacial tension, which acts so as to make the interface as small as possible, imparts a certain stiffness to the interface. The interface can be softened by introducing interfacially active molecules, e.g., amphiphiles like lipids, which accumulate in the interface and lower the interfacial tension. If there is a sufficient amount of amphiphiles, the interface can be fully covered, as shown in Fig. 5.1a and c. This implies that the area is essentially fixed and the interfacial tension tends toward zero. In that case, which is true for many fluid membranes, the stability and conformation of the interface is controlled by conformational entropy and the elasto-mechanical properties of the interface.

An interface can be considered soft in several ways. It can be easy to bend, as illustrated in Fig. 5.3a; it can be easy to compress or expand, as illustrated in Fig. 5.3b; or it can be easy to shear, as illustrated in Fig. 5.3c. In the case of a fluid interface, which is the case pertaining to the lipid bilayer of a biological membrane, the resistance to shearing is nil, and we can neglect that mode. The two other ways of deforming the interface are associated with two parameters, two elasto-mechanical modules, termed the bending modulus, κ, and the area compressibility modulus, K, respectively. The area compressibility modulus is defined via the energy per unit area, E_K, one has to spend in order to uniformly stretch an interface of area $A_\circ$, to produce an area change of ΔA according to the well-known Hooke's law for an elastic spring:

$$E_K = \frac{1}{2} K \left(\frac{\Delta A}{A_\circ} \right)^2 . \tag{5.2}$$

The bending modulus for a flat interface (we shall in Sect. 6.1 describe the case of curved interfaces) is defined via the energy per unit area, E_κ, that is required to produce a mean curvature, H, of the interface according to

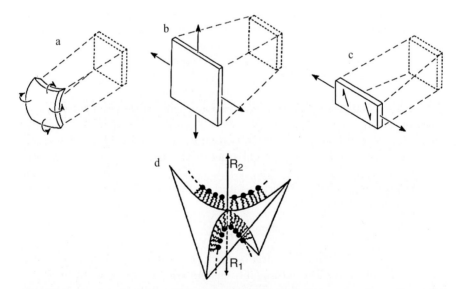

Fig. 5.3. Bending (**a**) and stretching (or compressing) (**b**), or shearing (**c**) a soft interface like a membrane. The curvature of an interface is characterized geometrically by the two radii of curvature, R_1 and R_2, indicated in (**d**)

$$E_\kappa = 2\kappa H^2 ,\qquad(5.3)$$

where the mean curvature is given by the two principal radii of curvature, R_1 and R_2, defined in Fig. 5.3d, as

$$H = \frac{1}{2}\left(\frac{1}{R_1} + \frac{1}{R_2}\right) .\qquad(5.4)$$

In the definition of the bending energy, E_κ, we have assumed that there is no internal structure of the interface and that there are no constraints imposed by boundaries, i.e., that the interface has to close onto itself. We shall come back to this constraint in Chap. 6. Obviously, the two modules κ and K must be related. It can be shown that this relation in the simplest case can be written as $\kappa = d_L^2 K$, where d_L is the thickness of the interface.

In Fig. 5.4 are shown the contours of two closed soft interfaces, a giant liposome and a red blood cell membrane. Both of these bodies are soft, but their ability to bend is different. The length scale over which they appear flat and smooth, i.e., the persistence length, ξ, is different. The persistence length is indeed related to the bending modulus via the relation

$$\xi \sim \exp\left(\frac{c\kappa}{k_B T}\right) ,\qquad(5.5)$$

where c is a constant. According to this equation, it is the ratio between the bending modulus and the thermal energy, $k_B T$, that determines the persistence length. Hence, ξ depends exponentially on the bending modulus.

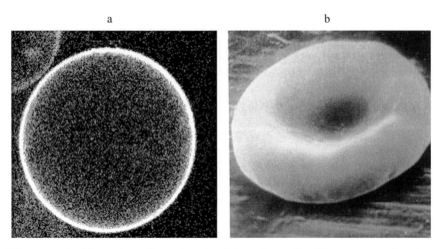

Fig. 5.4a–b. Examples of two closed soft interfaces. (**a**) The contour of a giant liposome of 60 μm diameter imaged by fluorescence microscopy. (**b**) A red blood cell of 5 μm diameter

Liposomal membranes can easily be prepared to be very soft and exhibit low values of κ, as described in Sect. 5.4 below. This allows for very substantial thermally driven fluctuations and surface undulations. Plasma membranes of cells usually have bending modules that are considerably larger than the thermal energy, $\kappa \gg k_B T$, so the persistence length for the membrane is larger than the size of the object. Hence, the plasma membrane in Fig. 5.4b appears smooth. It is interesting to note that some internal membranes in eukaryotic cells, e.g., Golgi and endoplasmic reticulum, as shown in Fig. 1.5, appear to be very soft and strongly convoluted, sometimes exhibiting nonspherical topologies. As we shall discuss in Sect. 9.4, the absence of cholesterol in these membranes may partly explain their apparent softness. Cholesterol tends to increase the value of the bending rigidity κ.

5.3 Forces Between Soft Interfaces

The softness of interfaces in general and membranes in particular has some striking consequences for the colloidal forces that act between them. A colloidal force is a thermodynamic force. In contrast to a mechanical force, which is determined by the gradient of a mechanical energy (or enthalpy, H) the colloidal force, F, is a spatial derivative of a free energy, $G = H - TS$, given by

$$F = -\left(\frac{\partial G}{\partial r}\right) = -\left(\frac{\partial H}{\partial r}\right) + T\left(\frac{\partial S}{\partial r}\right), \tag{5.6}$$

where r is the distance between the objects. Hence, the colloidal force involves the entropy, S. This implies that there is always an entropic repulsion between soft interfaces, even in the extreme case (like an ideal gas) where there are no direct mechanical forces in effect and the first term of the right-hand side of (5.6) tends to zero. It is the reduction in configurational entropy due to the confinement that produces the repulsive force.

Several examples of this scenario are illustrated in Fig. 5.5. The extreme case is that of a micro-emulsion, which is basically a dense gas of very soft, strongly repelling and fluctuating interfaces, as shown in Fig. 5.5f. A special version of a colloidal particle covered by polymers is a liposome incorporated with special lipids, so-called lipopolymers, to whose head groups are attached polymer chains. Such liposomes are called *stealth liposomes* for reasons that will be revealed in Sect. 20.3. These liposomes can be used as drug carriers to circumvent the immune system, partly because of the entropic repulsion that results from the softness of the polymer cushion on their surface.

An explicit expression for the entropic *undulation force* (sometimes referred to as an osmotic pressure) acting between a stack of soft interfaces like in Fig. 5.5e with spacing d was derived by the German physicist Wolfgang Helfrich to be

$$F \sim \frac{(k_B T)^2}{\kappa d^3}. \tag{5.7}$$

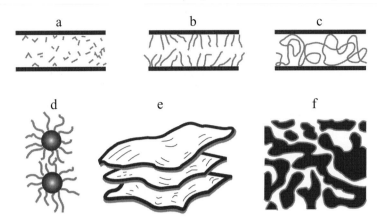

Fig. 5.5a–f. Examples of systems in which entropic factors lead to repulsive forces, cf. (5.6). (**a**) Ideal gas between two hard walls of a container. (**b**) Hard surfaces grafted with long-chain molecules, e.g., polymers. (**c**) Polymers confined between hard surfaces. (**d**) Colloidal particles coated by flexible polymers. (**e**) Stack of soft interfaces, e.g., membranes. (**f**) Micro-emulsion

According to this equation, the repulsive force increases as the bending rigidity, κ, is diminished. This has important consequences for the interaction between lipid membranes, as shown in Sect. 5.4 below.

5.4 Lipid Membranes are Really Soft

A visual impression of the softness of a lipid bilayer in the form of a unilamellar liposome can be obtained from Fig. 5.6a, which shows a series of contours of a giant liposome observed in a microscope at different times. A substantial variation in the contour is seen over time. This is a manifestation of thermally induced surface fluctuations or undulations. The intensity of the fluctuations, considering the size of the liposome in relation to the thickness of the bilayer, shows that lipid membranes are really soft. The softness in terms of the bending modulus, κ, can be extracted from an analysis of the spectrum of fluctuations.

The softness of a lipid bilayer in terms of its area compressibility modulus, K, can be studied using micro-mechanical techniques. Using a glass pipette with a very small diameter, of about 1 to 10 µm, it is possible by aspiration to apply a stress, τ, to the membrane and subsequently measure the resulting area strain, $\Delta A/A_\circ$, as illustrated in Fig. 5.6b, simply by measuring the expansion of the membrane into the pipette. Area strains of up to a couple of percent can typically be obtained before the membrane is broken. K can subsequently be determined from (5.2), since $\tau = K(\Delta A/A_\circ)$.

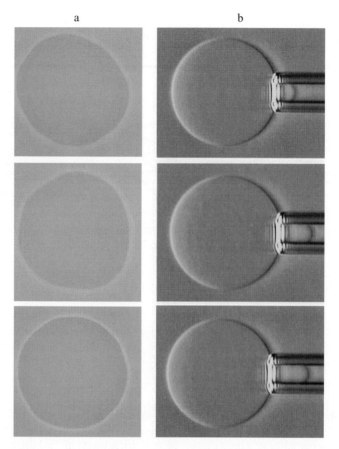

Fig. 5.6. (a) Fluctuating liposomes of diameters around 50 μm. (b) Aspiration of a giant vesicle of diameter 32 μm into a micro-pipette. Suction pressure is increased from *top* to *bottom*, leading to a change in vesicle area

To get a feeling of the softness of natural materials in the form of membranes and lipid bilayers it is instructive to compare it to the softness of a man-made material, e.g., a simple soft plastic like polyethylene. When it comes to bending, a DMPC lipid bilayer is about five times softer than a red blood cell membrane, which, in turn, is 50,000 times softer than a film of polyethylene of the same thickness. No wonder a closed bag of polyethylene cannot do what a red blood cell must do in its life span of 120 days in circulation: it travels 400 km, and during its excursions into the fine and narrow blood capillaries it has to stretch and bend to change its shape by a very large amount, more than 100,000 times without falling apart. In terms of the area compressibility, a DMPC bilayer is about ten times softer than a red blood cell membrane, which, in turn, is about five times softer than a film of polyethylene of the same thickness.

5.4 Lipid Membranes are Really Soft

The reason why the membrane of the red blood cell is less soft than a DMPC bilayer is that the red cell has a *cytoskeleton*. If the lipids are extracted from the red cell membrane and re-formed as a lipid bilayer, this bilayer is considerably softer but is still less soft than DMPC bilayers. The reason for this is that the red blood cell, being an eukaryote (although without a cell nucleus), has a plasma membrane that contains large amounts of cholesterol, typically 30%. Cholesterol tends to make membranes less soft, both in terms of bending stiffness and area compressibility. Other factors that influence the softness are fatty-acid chain length and degree of saturation of the chains. The general trend is that shorter and more unsaturated chains provide for greater softness. Furthermore, various solutes can influence the softness quite dramatically. The typical values of the elasto-mechanical modules for lipid bilayers correspond to energies that are in the range of the thermal energy, $k_B T$, e.g., κ for DMPC is around $10\, k_B T$. Hence, the elastic membrane fluctuations are expected to be very sensitive to temperature. This will have some dramatic consequences at membrane phase transitions, as discussed in Chap. 12.

A conspicuous consequence of undulation forces experienced by bilayers that become soft is that a lipid bilayer or a membrane that is adsorbed to a solid surface should be repelled from the surface if the bilayer is softened. This effect, which may be of importance for cell-cell adhesion and possibly for the motility of unicellular organisms, has indeed been observed. Two examples are illustrated schematically in Fig. 5.7. Figure 5.7a shows soft lipid bilayers that are being repelled from a hard surface and from each other by undulation forces, leading to unbinding. Figure 5.7b shows a vesicle or cell-like object that, due to renormalization of its bending modulus, is made to hop off the surface to which it adheres.

The strong effects on the softening of lipid bilayers discussed above is due to fluctuations in density. Bilayers can also be made softer by compositional fluctuations. In this case, the softening is due to local variations in the

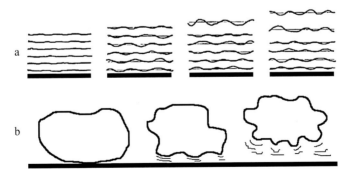

Fig. 5.7. (a) Lifting off bilayers from a stack by undulation forces. (b) Lifting off a vesicle or cell from a surface by undulation forces

composition. Close to so-called critical mixing points, the membrane composition can fluctuate strongly, as described in Sect. 9.3. Hence, at any given time, the local composition at a given place in the bilayer can be very different from the average global composition of the bilayer. We shall have a closer look at this phenomenon in Sect. 11.2. Both fluctuations in density and composition tend to lower the bending modulus, κ.

One of the major questions we shall address in the following chapters is the microscopic and molecular origin of membrane softness and how it is manifested in membrane structure on the nanometer scale. This may provide some clues as to how the softness eventually can be controlled. It is the hypothesis that the lipid-bilayer softness, the dynamic structure of the membrane, and the corresponding lipid organization are important regulators of membrane function and the ability of the membrane to support biological activity. A consequence of this hypothesis is that the generic effects of peptides, proteins, and drugs on membrane structure and function, on the one side, and the influence of bilayer structure on these compounds, on the other side, may be understood in part by the ability of these compounds to alter lipid-bilayer softness and molecular organization.

6 Soft Shells Shape Up

6.1 Bending Interfaces

It is instructive to consider the spatial dimensions of a membrane system like a uni-lamellar vesicle, as shown in Fig. 3.5. Whereas the lipid bilayer itself is only about 5 nm thick, the diameters of vesicles and liposomes are orders of magnitude larger, typically in the range of 50 nm to 50,000 nm. Hence, lipid bilayers are extremely thin films of tremendous anisotropy. It is therefore to be expected that some of the generic properties of vesicles and possibly cells can be understood and described by considering the membranes as infinitely thin shells associated with unique material characteristics. Many theoretical approaches to determine membrane conformations, topology, and shapes therefore assume the membrane to be a two-dimensional liquid interface imbedded into a three-dimensional space.

A liquid interface exhibits no resistance to shearing. Therefore, when it is mechanically deformed, there are only the two possible modes of deformation illustrated in Fig. 5.3a and b: bending and stretching/compressing. If we for a moment assume that the interface is infinitely thin with no internal structure and its area furthermore is fixed, we can neglect area compressibility and are then left with the sole possibility of bending. Bending leads to curvature, and the curvature of the interface at any given point is described by two principal radii of curvature, R_1 and R_2, as illustrated in Fig. 6.1.

In order to determine the most likely shape of an interface that is left alone, we need an expression for the energy that is involved in the bending. The German physicist Wolfgang Helfrich formulated in 1973 the most general expression for the elastic bending energy, dE_{surface}, that is required for deforming an element of area dA to a shape described by R_1 and R_2 (cf. (5.3) and (5.4))

$$dE_{\text{surface}} = \left[\frac{\kappa}{2} \left(\frac{1}{R_1} + \frac{1}{R_2} \right)^2 + \frac{\kappa_G}{R_1 R_2} \right] dA . \tag{6.1}$$

The total bending energy for deforming the entire interface called S is then the sum (i.e., the integral) over all the elements of area, i.e.,

$$E_{\text{surface}} = \oint_S dE_{\text{surface}} = E_\kappa + E_{\kappa_G} . \tag{6.2}$$

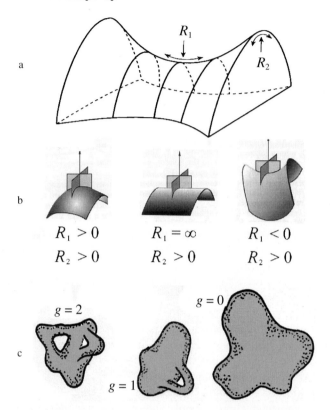

Fig. 6.1. (a) The two principal radii of curvature describing the local curvature of a mathematical interface. (b) From *left* to *right*: "Sphere", where R_1 and R_2 have the same sign, in this case positive. The mean curvature, $H = 1/2\,(1/R_1 + 1/R_2)$, is nonzero. "Cylinder", where $R_1 = \infty$ and $R_2 > 0$. The mean curvature is nonzero. "Saddle", where R_1 and R_2 have opposite sign. For the special case of a minimal surface, $R_1 = -R_2$, and the mean curvature is zero. (c) Closed interfaces of different topology characterized by different values of the genus number g

From these expressions we see that, in addition to the normal bending modulus κ (also called the mean curvature bending modulus), which we considered in Sect. 5.2 and in (5.3), there is an additional property of the interface, κ_G, that will determine the bending energy and hence the shape of the interface. κ_G is the so-called Gaussian curvature modulus (or the saddle-splay modulus).

The Gaussian curvature modulus, κ_G, controls the topological complexity of the interface. This is most easily seen by noting that for a closed interface, the contribution from the Gaussian curvature to the energy of the interface is proportional to a topological invariant, $4\pi(1-g)$, according to

$$E_{\kappa_G} = \kappa_G \oint_S \frac{1}{R_1 R_2} dA = \kappa_G 4\pi(1-g) \,. \tag{6.3}$$

This relationship is purely mathematical and is known as the Gauss-Bonnet theorem. g is the so-called genus number, which describes the topology of the closed interface, as illustrated in Fig. 6.1c. For a sphere, $g = 0$. More complex surfaces with holes have higher values of g. The value of the Gaussian curvature modulus is difficult to determine experimentally. κ_G is generally believed to be of the same order of magnitude as the mean curvature modulus, κ.

In order to get an intuitive feeling about the contribution of these two conceptually very different terms to the bending energy of an interface, let us consider a special class of interfaces or surfaces that are called minimal surfaces. Minimal surfaces have zero mean curvature everywhere, i.e, $H = 0$. A flat lamellar interface and the interface defined by the cubic structure in Fig. 6.2 are examples of minimal surfaces. For lamellar and cubic structures, which are perfectly ordered at very low temperatures, as shown to the left in Fig. 6.2, the mean curvature is identically zero, $H = 1/2\left(1/R_1 + 1/R_2\right) = 0$, and $E_\kappa = 0$. For the lamellar structure, $E_{\kappa_G} = 0$, whereas it is different from zero for the cubic structure. Obviously, the cubic structure, which is a bunch

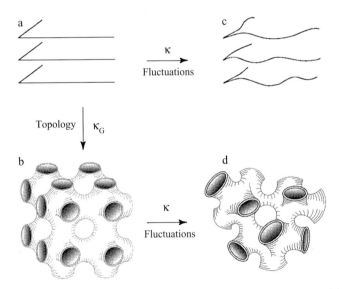

Fig. 6.2. Minimal surfaces with zero mean curvature: the lamellar (**a**) and the cubic structure (**b**). The surface structures have perfect order corresponding to low temperatures. Fluctuations at elevated temperatures lead to the fluctuating lamellar structure in (**c**) and the sponge structure in (**d**), respectively. The associated elastic bending energy is related to the bending modulus, κ. For negative values of the Gaussian curvature modulus, κ_G, the lamellar structure becomes unstable toward formation of the ordered cubic structure and the sponge structure, which have a more complex topology than the lamellar structure

of saddles, will be stabilized for large negative values of κ_G, whereas the lamellar structure will be stabilized for large positive values of κ.

When fluctuations are introduced, e.g., by increasing the temperature, both the lamellar and the cubic structure will be able to assume some local mean curvature. As shown to the right in Fig. 6.2, this leads to undulations on the lamellar interface and possibly to a disordering of the cubic structure into a sponge structure. The mean curvature modulus, κ, serves to control the amplitude of the thermal fluctuations by assuring that the deviation from the zero mean curvature is as small as possible.

6.2 Spontaneous Curvature

In the description of the bending of fluid interfaces in Sect. 6.1 above, we assumed that the interfaces had no internal structure and were infinitely thin. For real interfaces like lipid monolayers and bilayers, the internal structure of the interface and its thickness have to be taken into account. As described in Sects. 4.2 and 4.3 and illustrated in Fig. 4.6, lipid monolayers can exhibit *spontaneous curvature* due to the fact that lipid molecules have shape.

When two symmetric lipid monolayers have to live together in a bilayer, the bilayer itself has no intrinsic tendency to curve and its spontaneous curvature, C_0, is zero. However, if the two monolayers are chemically different, the bilayer prefers to curve and has a finite spontaneous curvature $C_0 = R_0^{-1}$. As we shall see shortly, there are several other possibilities of inducing an asymmetric condition for the bilayer to induce a spontaneous curvature. The description of the bending energy of a membrane has to take into account that the bilayer can have an intrinsic desire to bend when left alone. This is readily done by rewriting (6.1) as

$$dE_{\text{surface}} = \left[\frac{\kappa}{2} \left(\frac{1}{R_1} + \frac{1}{R_2} - \frac{2}{R_0} \right)^2 + \frac{\kappa_G}{R_1 R_2} \right] dA . \tag{6.4}$$

For fixed topology, the bending energy will then be minimal when the bilayer assumes a curvature corresponding to its spontaneous curvature. We shall in the following restrict ourselves to considering only membranes of spherical topology.

There are a number of different sources for spontaneous curvature in lipid bilayers and biological membranes in addition to the simple one due to a possible chemical difference between the two lipid monolayers. Figure 6.3 gives a schematic presentation of a closed membrane vesicle that is subject to various asymmetric conditions leading to a nonzero spontaneous curvature. Some of these conditions can be different for different vesicles in the same preparation, even if they are made of the same lipids. The intrinsic spontaneous curvature can also vary from vesicle to vesicle, and it is usually difficult to measure experimentally.

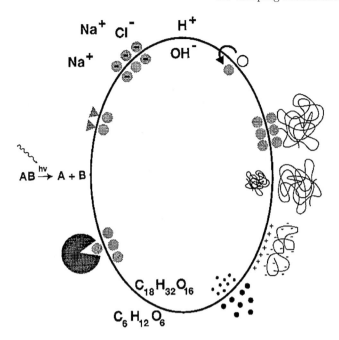

Fig. 6.3. Schematic illustration of a closed membrane that becomes asymmetric by a number of chemical, physical, and functional mechanisms. The asymmetry leads to a nonzero spontaneous curvature, $C_0 = R_0^{-1}$.

6.3 Shaping Membranes

Before we are ready to discuss shapes of closed membranes we have to mention another circumstance that can make it necessary to consider different vesicles independently. It has to do with a fairly obvious but often overlooked fact. A lipid membrane in the form of a vesicle has a history in the sense that it is made of a lipid bilayer which, in order to avoid problems with an open boundary, at some stage closes into a closed shape. This implies that the initially unstressed area of the inner monolayer, A^{inner}, and the initially unstressed area of the outer monolayer, A_o^{outer}, have to compress or stretch in order to produce a closed vesicle with a fixed mean area of $A = \frac{1}{2}(A^{\text{inner}} + A^{\text{outer}}) = \frac{1}{2}(A_o^{\text{inner}} + A_o^{\text{outer}})$. This implies that, although the mean area is fixed, there can be deviations in the differential area, $\Delta A = A^{\text{inner}} - A^{\text{outer}}$, from its equilibrium value, $\Delta A_o = A_o^{\text{inner}} - A_o^{\text{outer}}$. Stressing the two monolayers introduces an extra term in the total energy for the membrane, which includes the area compressibility modulus, κ, and the bilayer thickness, d_L. This so-called Area-Difference-Energy (ADE) term, which was first anticipated by the Canadian biophysicist Ling Miao and her collaborators, has the form

$$E_{\text{ADE}} = \frac{\alpha \kappa \pi}{2 A d_L^2}(\Delta A - \Delta A_\circ)^2 \ , \qquad (6.5)$$

where α is a constant that is close to unity for all phospholipids. In contrast to the spontaneous curvature, the differential area is not an intrinsic property of the bilayer but set by the way the closed bilayer membrane happens to be prepared.

The total bending energy, which will determine the equilibrium shape of a closed lipid vesicle, is then given by

$$E_{\text{total}} = \frac{\kappa}{2} \oint_S \left(\frac{1}{R_1} + \frac{1}{R_2} - \frac{2}{R_0} \right)^2 dA + \frac{\alpha \kappa \pi}{2 A d_L^2}(\Delta A - \Delta A_\circ)^2 \ . \qquad (6.6)$$

The equilibrium shape of a given fluid vesicle is obtained by minimizing E_{total}, given the geometrical parameters of the vesicle, i.e., area A and volume V, and the materials parameters, i.e., spontaneous curvature R_0^{-1} and preferred differential area $\Delta A_\circ$. The effects of temperature are neglected here. This is a reasonable procedure since the bending modulus is typically ten times the thermal energy, $\kappa \sim 10 \, k_B T$. The results for the equilibrium shapes are given in the form of a phase diagram as shown in Fig. 6.4.

The phase diagram exhibits an enormous richness of shapes. The theoretically predicted shapes as well as the transitions between different shapes are generally in good agreement with findings in the laboratory. In Fig. 6.5 are shown a series of vesicles with different shapes. For each type of shape, the vesicles will under experimental circumstances exhibit thermally induced fluctuations in shape.

When varying the parameters that determine the shape according to the phase diagram, transitions from one type of shape to another can be observed. This is most easily done by varying the osmotic pressure across the membrane, e.g., by changing the sugar solution on the outside. The transition will take some time, since it takes some time for water to diffuse across the bilayer to establish the new equilibrium. An example of the transition from one shape to another is illustrated in Fig. 6.6. This figure shows a time sequence of the transition across the line from pear-shaped vesicles to a vesicle with a bud, cf. the phase diagram in Fig. 6.4. This transition is called the *budding transition*.

6.4 Red Blood Cells Shape Up

Red blood cells have plasma membranes that are much more complex than the lipid membranes we have considered so far. In addition to the fluid lipid bilayer, the red blood cell membrane has a cytoskeleton that is a cross-linked network of the protein spectrin. Therefore, bending and changing the shape of a red blood cell membrane involve both deformations of the lipid bilayer as well as elastic deformation of the cytoskeleton, which is a tethered membrane

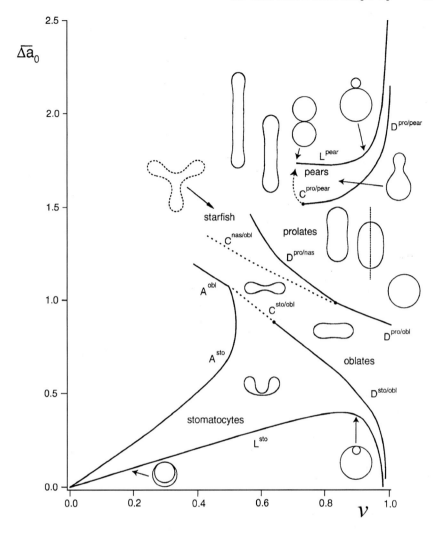

Fig. 6.4. Phase diagram predicted for closed vesicle membranes. Stable shapes are given as a function of two reduced parameters: the reduced vesicle volume, v, where $v = 1$ corresponds to a sphere, and a parameter, $\overline{\Delta a_o}$, which is a combination of spontaneous curvature and preferred differential area. $\overline{\Delta a_o}$ is a measure of the preferred curvature of the vesicle. The names of several of the shapes refer to the shapes that have been found for red blood cells

with a certain resistance to shearing, as illustrated in Fig. 5.2b and c. Our understanding of the relationship between the morphology of lipid bilayer vesicles described above and the shapes of red blood cells is still very premature. However, some of the gross aspects of the red blood cell shapes indicate

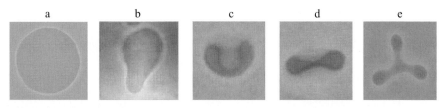

Fig. 6.5a–e. Gallery of lipid vesicle shapes found experimentally. (**a**) Sphere. (**b**) Pear shape. (**c**) Stomatocyte. (**d**) Discocyte. (**e**) Starfish shape

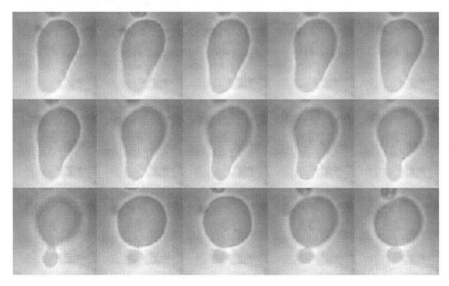

Fig. 6.6. Series of snapshots, 1.2 sec apart, of the dynamic change of shape of a vesicle from a pear shape to a spherical vesicle with a bud. Time lapses from *top left* to *bottom right*

that they are governed by a similar physics related to bending elasticity as simple lipid vesicles are.

A healthy blood cell has a biconcave shape, a so-called discocyte, as illustrated in Fig. 6.7. It has been known for many years that this shape can be turned into other shapes by the influence of a number of factors such as osmotic pressure, cholesterol content, pH, level of ATP, as well as various other externally added amphiphatic compounds. Some of these other shapes are characteristic of various diseases. Two shapes are of particular interest, the so-called stomatocyte, shown in the left-hand side of Fig. 6.7, and the echinocyte, shown in right-hand side of Fig. 6.7. The normal discocyte has a shape in between these two extremes. The curious thing is that when the lipids are extracted from the cell membrane, the shape of the spectrin network becomes almost spherical. Hence, the precise red blood cell shape is

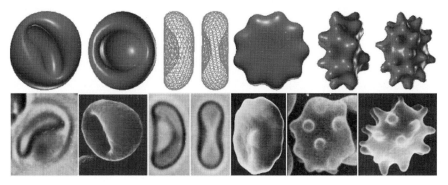

Fig. 6.7. Human red blood cells of varying shape, going from the stomatocyte (*left*), over the normal discocyte, to the spiked echinocyte (*right*). The *bottom* panel shows experimental shapes and the *top* panel shows theoretically predicted shapes

a manifestation of a coupling between the lipid bilayer properties and the elastic properties of the cytoskeleton.

The Canadian physicist Michael Wortis and his collaborators have shown that the sequence of human red blood cell shapes, from the stomatocyte, over the normal discocyte, to the spiked echinocyte, can be described theoretically by the ADE-model described in Sect. 6.3 above when including the stretch and shear elasticity properties of the protein-based cytoskeleton. The correlation between the experimental shapes and the theoretical predictions is surprisingly good, as seen by comparing the two panels in Fig. 6.7. Within the model, a single parameter describes the transition between the different shapes. This parameter is related to the preferred differential area ΔA_o in (6.6), which is a measure of the relaxed area difference between the inner and the outer monolayer leaflets of the lipid bilayer membrane. The beauty of this model lies in the fact that the red blood cell shapes can be described by a general physical model and the detailed biochemistry only enters in the way it determines the actual value of ΔA_o. As an example, it is known that cholesterol is predominantly incorporated in the outer leaflet, leading to an increase in ΔA_o, thereby shifting the shape towards the echinocyte. This is consistent with experimental observations. Similarly, cholesterol depletion of red cells leads to stomatocytes, which correlates to a lowering of the preferred differential area.

7 Biological Membranes – Models and Fashion

7.1 What Is a Model?

A *model* is an abstraction of nature. It can be very concrete and practical, e.g., given by a protocol to construct a particular sample for experimental investigation, or it can be given by a precise mathematical formula that lends itself to a theoretical calculation. A model can also be less well-defined and sometimes even implicit in the mind of the researcher. The concept of a model or *model system* is one of the cornerstones in natural sciences. It is a powerful and necessary tool to facilitate our perception of complex natural phenomena. The model helps us ask some relevant and fruitful questions out of the millions of possible questions that can be asked. It helps us guide experiments and perform theoretical calculations. And it is instrumental in interpreting the results of our endeavors. A good model is a blessing, but it can also be a curse. It may bias our thinking too strongly if we forget that it is just a model and not Nature herself. Models must constantly be scrutinized and questioned, even the most successful ones, not least because models reflect fashion among scientists.

A key element in the formulation of a useful model of such a complex system like a biological membrane is to strike the proper balance between general principles and specific details – or to balance the sometimes conflicting demands for truth and clarity. This can be illustrated by the photo of a ship in Fig. 7.1a. The construction blueprint in (b) seeks to capture as many details as possible of the ship, although it is clearly a model abstraction. Finally, the primitive drawing in (c) of the ship is a very simplified model of a ship. It lacks all sorts of details and is clearly out of proportion. Nevertheless, any kid can tell that this is a ship that can sail on water and is driven by some kind of motor. Depending on what you need, you would choose one of the three representations. They each have their virtues, even the hand-drawn one. You may want to choose that one if you are in the process of investigating a ship that you only know a little about. It contains very little bias and few details. Too many details will render the model applicable only to specific cases, and the details may obscure the generic underlying principles of organization. On the other hand, a too general model may provide little mechanistic insight, which makes the model less useful for the design of further and more penetrating critical investigations.

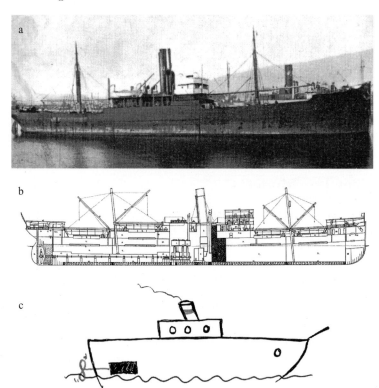

Fig. 7.1a–c. Ships and models of ships. (**a**) A real ship – that is, a photograph of a real ship. (**b**) A construction blueprint of a ship. (**c**) The author's drawing of a ship

In the case of biological membranes, the important elements of a model are likely to depend on which length and time scales are relevant for describing the problem of interest. This can imply serious difficulties since many membrane properties are controlled by phenomena that take place over a wide range of scales that are mutually coupled. It is likely that one will be best served by working with a set of membrane models, experimental as well as theoretical ones, chosen according to the particular type of question under consideration – and with due reference to which time and length scales are expected to be relevant.

7.2 Brief History of Membrane Models

In 1972, Singer and Nicolson proposed their celebrated *fluid-mosaic model* of biological membranes. The Singer-Nicolson model has since been a central paradigm in membrane science. The simple yet powerful conceptual framework it provided continues to have an enormous impact on the field

7.2 Brief History of Membrane Models

of membranes. As a key property, the Singer-Nicolson model assigned to the lipid-bilayer component of membranes a certain degree of *fluidity*. The fluidity concept was meant to characterize the lipid bilayer as a kind of pseudo two-dimensional liquid in which both lipids and membrane-associated proteins display sufficient lateral mobility in order to allow for function. The overall random appearance of this lipid-protein fluid composite made the membrane look like a mosaic. Except in cases where sterols or unsaturated lipid chains might alter the bilayer "fluidity," the conspicuous diversity in the chemical structures of lipids, which is actively maintained by cells, had little significance in the model. This lipid diversity, together with the varying but characteristic lipid composition of different types of cells and organelles, has become an increasing puzzle, which is exacerbated by the enhanced understanding of the variation in physical properties among different lipids and lipid assemblies.

When Singer and Nicolson proposed the fluid-mosaic model in 1972, membrane modelling already had come a long way, as illustrated in Fig. 7.2. The first important step was taken in 1925 by Gorter and Grendel who showed that the membrane is very thin, being only two molecules thick, as shown in Fig. 7.2a. The experiment behind this remarkable and fundamental insight was surprisingly simple and elegant. Gorter and Grendel made a lipid extract of red blood cells whose surface area was known from microscopy. The lipids were spread on a water surface and compressed to produce a dense lipid monolayer as in Fig. 3.4a. When measuring the area of the resulting monolayer, it was found to be twice that of the surface of the red cells that provided the lipid extract. Hence, it could simply be concluded that the membranes of the cells were only two molecules thick.

The association of membrane proteins with the lipid bilayer was introduced in the Danielli-Davson model. In this model, the proteins appeared as a kind of spread on the lipid polar head groups at the two sides of the lipid bilayer, as illustrated in Fig. 7.2b. A related version of membrane organization appears in Robertson's unit membrane model in which the proteins are pictured as stratified layers sandwiching the lipid bilayer, as shown in Fig. 7.2c. In the Singer-Nicolson fluid-mosaic model pictured in Fig. 7.2d, the proteins are grouped into two classes: integral membrane proteins that traverse the bilayer and primarily interact with the bilayer through hydrophobic forces, and peripheral membrane proteins that are peripherally associated with the lipid bilayer and primarily interact with the bilayer through polar (electrostatic and hydrogen bond) interactions. In either case, the proteins "float in a fluid sea."

Refinements of the fluid-mosaic model have been suggested from time to time, usually inspired by new insights obtained by focusing on some specific, or specialized, membrane feature. One example is the model by Jacob Israelachvili, who refined the Singer-Nicolson model to account for the need of membrane proteins and lipids to adjust to each other. This refined model also

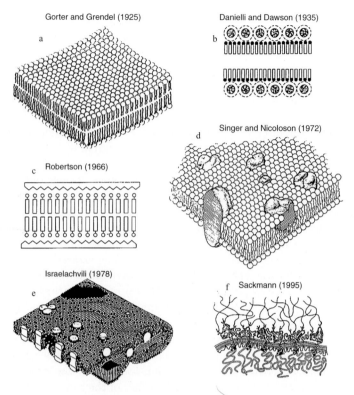

Fig. 7.2. Historic picture gallery of membrane modelling

incorporated membrane folding, pore formation, and thickness variations, as well as some degree of heterogeneity, as shown in Fig. 7.2e. Another elaboration of the Singer-Nicolson model, which emphasized the importance of the cytoskeleton and the glycocalyx, was developed by Erich Sackmann and is presented in Fig. 7.2f.

The various refinements of the Singer-Nicolson model represent the fashions in the field of membranes where researchers who investigate certain aspects of membrane complexity are in need of simple and transparent working models that can help them guide their intuition and facilitate the interpretation of experiments.

7.3 Do We Need a New Membrane Model?

There are several reasons to expect that we need a new model of biological membranes. Many of these reasons are dealt with in the present book. The notion of membrane fluidity, which was embodied in the Singer-Nicolson fluid-mosaic model in Fig. 7.2d, was important because it served to emphasize that membranes are dynamic structures. Unfortunately, many subsequent investigators assumed, explicitly or implicitly, that fluidity implies randomness. This assumption neglects that fluids or liquids may be structured on length scales in the nanometer range, which are difficult to access experimentally, as described in Chap. 11. Also, structuring in time, in particular the correlated dynamical phenomena characteristic of liquid crystals, was not appreciated as being important for membrane function in the Singer-Nicolson model.

However, lively dynamics is perhaps the most conspicuous feature of a liquid membrane (Sect. 8.4). The dynamics does not necessarily imply randomness and disorder. In fact, the many-body nature inherent in the molecular assembly of a membrane insures that local order and structure develop naturally from an initially disordered liquid. Finally, the fluid-mosaic model pictured the membrane as a flat, pseudo two-dimensional layer. This may be an artistic simplification. It nevertheless de-emphasizes the transverse dynamical modes of individual lipid molecules, as well as the existence of large-scale excursions into the third dimension with the ensuing curvature-stress fields (Chap. 8), instabilities toward non-lamellar symmetries (Chap. 4), and coupling between internal membrane structure and molecular organization (Chap. 15) on the one hand and membrane shape and shape transformations on the other (Chap. 6). All these phenomena are intimately related to the fact that membranes are pieces of soft condensed matter, as we saw in Chap. 5.

It is now recognized that the randomness implied in the fluid-mosaic membrane model does not exist. This recognition builds on a wealth of experimental results that show that the lateral distribution of molecular components in membranes is heterogeneous, both statically and dynamically – corresponding to an organization into compositionally distinct domains and compartments, as described in Chap. 11. In addition to immobilization and domain formation due to interactions between the cytoskeleton or the extracellular matrix and the membrane, several physical mechanisms generate dynamic lateral heterogeneity of both lipids and proteins in liquid membranes.

This nonrandom organization imposed by the fluid membrane means that membrane functions do not need to depend on random collisions and interactions among reactants, but may be steered in a well-defined manner that allows for a considerable mobility of the individual constituents. This dynamic organization of the membrane makes it sensitive to perturbations by both physical (e.g., temperature and pressure) as well as chemical (e.g., drugs and metabolites) factors (Chap. 17), which thus provides an exceptional vehicle for biological triggering and signalling processes (Chap. 15).

It has been suggested that the Singer-Nicolson model of membranes has been successful because it does not say (too) much. It does not bias the user strongly and hence allows for broad interpretations of new experimental data and novel theoretical concepts. This is the strength of the model. It is also its weakness, as it in many cases is not very helpful when questions are asked about membrane structure and, in particular, about membrane function. For those purposes, the model is too generic – in part because it provides too little, or no, insight into membrane protein assembly, lipid bilayer heterogeneity, monolayer or bilayer curvature, and bilayer bending and thickness fluctuations. Moreover, the model, by emphasizing stability, tends to de-emphasize dynamics; it does not address the issues relating to conformational transitions in membrane proteins and, just as importantly, the model does not address the conflict between the need for bilayer stability (the membrane must be a permeability barrier and consequently relatively defect-free) and the need for the bilayer to adapt to protein conformational changes. The bilayer must not be too stable because that would tend to limit protein dynamics. A manifestation of this dichotomy may be the widespread occurrence of lipids with the propensity for forming non-bilayer structures, as we discussed in Sects. 4.3 and 4.4.

7.4 Theoretical and Experimental Model Systems

Throughout this book we exploit the concept of models and modelling in our attempt to understand the role of lipids in membrane structure and functioning. In Chap. 6, we started out on the highest level of abstraction, where the membrane was considered a mathematical surface which is associated with mechanical properties and which displays complex conformations, shapes, and topologies. We found that the shapes of red blood cells could be understood and described within this general framework. The simplest experimental model of a lipid membrane will be dealt with in Chap. 10. Here we consider just half a membrane, a lipid monolayer at an air/water interface. Monolayer model membranes as well as the model bilayer membranes dealt with in Chaps. 8, 9, and 11 in the form of vesicles and liposomes are fairly easy to make since they basically self-assemble when dry powders of lipids are exposed to water. Some of the experimental bilayer models described in Chap. 11 are more complicated to manufacture since they require a solid support on which the layer is deposited.

The next level of complication in membrane modelling involves the incorporation of specific molecules in the lipid membranes, such as cholesterol dealt with in Sect. 9.4 and membrane proteins considered in Chap. 13. The experimental preparation of lipid-protein recombinants require special skills, in particular when functioning proteins and enzymes are involved.

We shall take advantage of theoretical models and model concepts throughout the presentation, ranging from very detailed microscopic models with atomistic detail, over various mesoscopic coarse-grained models where atomic and molecular details are hidden, to phenomenological models formulated in terms of macroscopic parameters. A very powerful tool exploited to derive the properties of the theoretical models is computer simulation techniques, using either stochastic principles, as in Monte Carlo simulation, or deterministic principles, as in Molecular Dynamics simulation.

Part II

Lipids Make Sense

8 Lipids in Bilayers – A Stressful and Busy Life

8.1 Trans-bilayer Structure

So far we considered membranes as ultrathin shells in which the internal membrane structure only entered very indirectly through the spontaneous curvature, through the observation that a bilayer consists of two monolayers and by the simple fact that the membrane has a thickness. In order to understand the organizational and functional principles of membranes, it is important to realize that a lipid bilayer is not just a homogeneous thin slap of a dielectric medium immersed in water but that the bilayer is a highly stratified structure with a distinct trans-bilayer molecular profile. This profile determines the membrane both as a barrier, carrier, and target. This is of particular importance for understanding how proteins function in and at membranes and how, for example, drugs interact with membranes.

The *trans-bilayer profile* is the most well-characterized structural property of bilayers since it most easily lends itself to being monitored by a number of techniques, e.g., X-ray and neutron-scattering techniques, magnetic resonance experiments, molecular-probe measurements, or computer simulation calculations. Magnetic resonance and molecular-probe techniques can give information about the structure and dynamics in various depths of the bilayer by using local reporter molecules or atoms. In addition to determining the thickness of membranes, scattering techniques can determine the probability of finding a specific part of a lipid molecule at a given depth in the bilayer. Computer simulation calculations can, if based on an accurate atomic-scale model with appropriate force fields, provide very detailed information on the structure and dynamics of bilayers.

Figure 8.1 illustrates two representations of the trans-bilayer structure of phospholipid bilayers. One representation (a) is a snapshot in time and the other (b) is an average over time. The fact that two representations are used illustrates the importance of realizing that lipid molecules in lipid bilayers are very lively, as we shall return to below.

Figure 8.1 shows the bilayer as a highly disordered liquid system with a distinct stratification. It can grossly be described in terms of four layers: (1) a layer of perturbed water, i.e., water that is structured and deprived of some of its hydrogen bonds, as described in Sect. 3.5, (2) a hydrophilic-hydrophobic region, including the lipid polar head groups as well

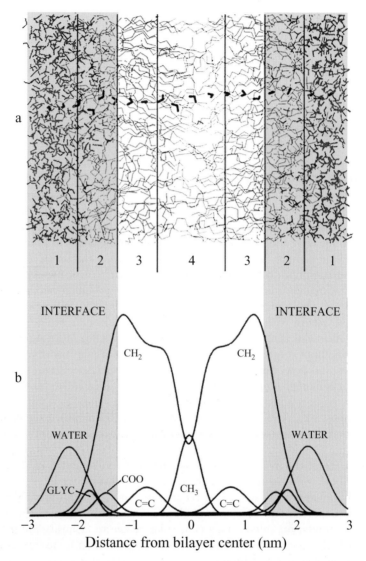

Fig. 8.1. (a) Trans-bilayer structure of a fluid DPPC lipid bilayer, as obtained from a Molecular Dynamics simulation. The four structurally different regions are labelled (1) perturbed water, (2) a hydrophilic-hydrophobic interfacial region involving the lipid polar-head groups, (3) a soft-polymer-like region of ordered fatty-acid chain segments, and (4) a hydrophobic core with disordered fatty-acid chain segments. A possible path for permeation of a single water molecule across the bilayer is indicated. (b) Trans-bilayer density profile of a fluid DOPC bilayer obtained from X-ray and neutron-scattering techniques. The curves give the relative probabilities of finding the different molecular segments of the phospholipid molecules. The highlighted region is the extended hydrophilic-hydrophobic interface

as both water and part of the upper segments of the fatty-acid chains, (3) a soft polymer-like region of ordered fatty-acid chain segments, and (4) a hydrophobic core with disordered fatty-acid chain segments of a structure similar to that of a liquid oil like decane. Although the detailed nature of such profiles depends on the actual lipid species in question, the overall structural stratification is generic for aqueous lipid bilayers.

The most striking and important observation to be made from Fig. 8.1 is that the region of space that makes up the hydrophobic-hydrophilic interface of the membrane, i.e., regions (1) and (2), occupies about half of the entire lipid-bilayer thickness. This is the extended hydrophilic-hydrophobic interface region highlighted in Fig. 8.1. The presence of this layer, its chemical heterogeneity as well as its dynamic nature, is probably the single most important quantitative piece of information on membrane structure and organization that has to be taken into account in the attempt to make a useful membrane model, as discussed in Chap. 7. It has been pointed out by the American biophysicist Steven White that the chemically heterogeneous nature of this extended interface region makes it prone to all sorts of noncovalent interactions with molecules, e.g., peptides and drugs, that bind, penetrate, and permeate membranes. This interface is thick enough to accommodate an α-helical peptide that lies parallel to the bilayer surface. This issue is further discussed in Chap. 17.

8.2 The Lateral Pressure Profile

Lipids in bilayers are kept in place because of the hydrophobic effect discussed in Sect. 3.4. This is a way to keep the oily fatty-acid chains away from the water. It is not an entirely happy situation for the lipid molecules, however. They are subject to large stresses by being confined in a bilayer structure along with their neighbors. In order to appreciate how stressful this can be, we have to examine the various forces that act inside the lipid bilayer. This will lead us to one of the most fundamental physical properties of lipid bilayers, the *lateral pressure profile*.

Figure 8.2a gives a schematic illustration of a cross section through a lipid bilayer indicating the forces that act to stabilize the layer. When the bilayer is in equilibrium, these forces have to sum up to zero. Since the forces, due to the finite thickness of the bilayer, operate in different planes, the pressures are distributed nonevenly across the bilayer, as shown schematically by the profile in Fig. 8.2b. This profile is called the lateral pressure or lateral stress profile of the bilayer.

The lateral pressure profile is built up from three contributions. A positive pressure resulting from the repulsive forces that act between the head groups, a negative pressure (the interfacial tension) that acts in the hydrophobic-hydrophilic interface as a result of the hydrophobic effect, and a positive pressure arising from the entropic repulsion between the flexible fatty-acid

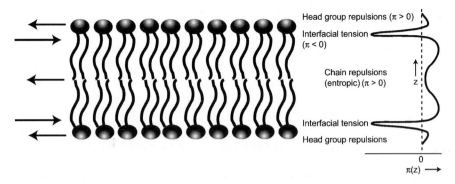

Fig. 8.2. Lateral pressure profile of a lipid bilayer. (**a**) Schematic illustration of a cross section through a symmetric lipid bilayer with an indication of the forces that act within the layer. (**b**) The resulting pressure or stress profile

chains (chain pressure). The detailed form of the pressure profile depends on the type of lipids under consideration. Due to the small thickness of the lipid bilayer, the rather large interfacial tension from the two interfaces of the bilayer has to be distributed over a very short range. This implies that the counteracting pressure from the fatty-acid chains has to have an enormous density, typically around several hundreds of atmospheres. This is easily seen by noting that the interfacial tension at each of the two hydrophobic-hydrophilic interfaces of a lipid bilayer is around $\gamma = 50 \, \text{mN/m}$. The lateral pressure of the interior of the lipid bilayer has to counterbalance this tension over a distance corresponding to the bilayer thickness, d_L, which is only about 2.5–3 nm. The lateral pressure density (force per unit area) of the bilayer then becomes $2\gamma/d_\text{L}$, which amounts on average to about 350 atm. Pressure densities of this magnitude are capable of influencing the molecular conformation of proteins imbedded in the membrane and hence provide a possible nonspecific coupling between the lipid membrane and the function of proteins, as we shall see in Sect. 15.2.

With reference to the discussion in Chap. 4 about the effective shape of lipid molecules, it is now clear from the description of the lateral pressure profile and Fig. 8.2 that it is not possible to assign a well-defined shape to a lipid molecule imbedded in a bilayer. The stressed and frustrated situation that a lipid molecule experiences in a bilayer is better described by the pressure profile, although there is no simple relation between the molecular structure and the actual distribution of stresses in the bilayer. Therefore, it is the lateral pressure profile that is the more fundamental physical property and that underlies the curvature stress field introduced in Sect. 4.3. It is therefore also the lateral pressure profile that determines bilayer spontaneous curvature, as well as the mean curvature and the Gaussian curvature modules described in Sects. 6.1 and 6.2.

8.3 How Thick Are Membranes?

Due to the fact that lipid bilayers are very stratified, the question of how thick a membrane is requires some qualification. Obviously, there are several different average thicknesses one can inquire about. Moreover, the various dynamical modes make it questionable what a thickness measure can be used for. In Sect. 13.3, we shall describe a fundamental principle for the interaction of trans-membrane proteins with lipid bilayers. This principle is based on matching the average hydrophobic thickness of the lipid bilayer and the hydrophobic length of the part of the protein that traverses the lipid bilayer. For this purpose we need an estimate of the hydrophobic thickness.

Obviously, the thickness of a lipid bilayer membrane depends on the length and degree of saturation of the fatty-acid chains of which its lipids are made. The longer the chains and the more saturated they are, the thicker the bilayer will be. The thickness also depends on the degree of hydration. The less hydrated the thicker the bilayer will be, because dehydration causes the head groups and therefore the fatty-acid chains to get closer together and hence stretch out. A very important determinant of lipid bilayer thickness is cholesterol. The reason for this, which is discussed in Chap. 9.4, is related to the fact that cholesterol has a strong tendency to stretch out and order the fatty-acid chains of the phospholipids. Hence, liquid lipid bilayers are usually thickened by cholesterol. Finally, temperature has a dramatic effect on lipid bilayer thickness: The higher the temperature, the thinner the bilayer. Under certain circumstances, to be discussed in Sect. 9.2, the lipid bilayer undergoes a phase transition, the so-called main phase transition. At this transition, the bilayer thickness can vary very abruptly.

The thickness of a lipid bilayer can be measured by X-ray or neutron scattering techniques applied to single bilayers on a solid support (cf. Fig. 11.4a) or to lamellar stacks of bilayers, as schematically represented in Fig. 8.3a. Results for the hydrophobic thickness as a function of temperature for DMPC and DPPC bilayers are shown in Fig. 8.3b. A dramatic reduction of thickness is observed as the bilayers are taken through their respective main phase transition. The thickness is seen to be larger for the lipid species with the longer chains. Moreover, the jump in thickness is more abrupt the longer the chains are. This systematics holds also for other chain lengths.

The hydrophobic membrane thickness of lipid bilayers in the liquid phase is strongly dependent on the amount of cholesterol incorporated into the bilayer. As an example, the thickness of a mono-unsaturated POPC lipid bilayer in its liquid phase can increase as much as 15–20% upon varying the cholesterol concentration from 0 to 30%, which is the level in most eukaryotic plasma membranes.

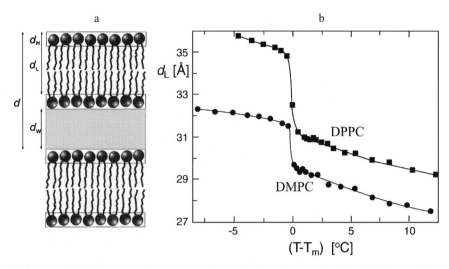

Fig. 8.3. (a) Schematic illustration of a multi-lamellar stack of lipid bilayers. The lamellar repeat distance is d, the hydrophobic lipid bilayer thickness is d_L, the thickness of a head group layer is d_H, and the water layer thickness is d_W. (b) Hydrophobic thickness, d_L, as a function of temperature for DMPC and DPPC lipid bilayers in the neighborhood of their respective main phase transition temperatures, T_m

8.4 Lively Lipids on the Move

Lipid molecules in liquid bilayers are extremely lively and undergo a range of different dynamical processes. They are constantly changing intra-molecular conformations, they are wobbling, they are protruding out of the layer, and they are moving around. Figure 8.4 illustrates schematically some of the motions that lively individual lipids perform. These motions range over an enormous time span, from picoseconds to hours. Conformational changes can be fast, since they involve rotations around C–C bonds, which typically take a few picoseconds. The rotation of the lipid molecules are also fast and occur on a time scale of nanoseconds, whereas lateral diffusion is in the range of tens of nanoseconds. A typical lipid will on average rotate once around its axis while it travels a distance corresponding to its own size. The wobbling of the fatty-acid chain, which leads to changes in its direction within the bilayer, is much slower, typically of the order of tens of milliseconds.

The fast lateral mobility of lipids in the plane of the membrane is a typical liquid property. Over time, lipids will be able to explore the entire lipid bilayer or membrane. For a typical cell size, a lipid molecule can travel across the cell membrane within less than half a minute. Lipid molecules furthermore undergo substantial excursions perpendicular to the membrane plane in the form of single-molecule protrusions that take place over time scales of tens of picoseconds. The motion of lipid molecules from one monolayer leaflet

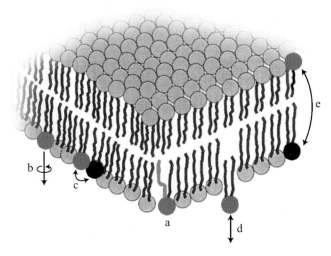

Fig. 8.4a–d. The many kinds of motions that lipid molecules in a lipid bilayer can perform. (**a**) Conformational change (see also Fig. 4.3). (**b**) Rotation around the molecular axis. (**c**) Lateral diffusion. (**d**) Protrusion out of bilayer plane. (**e**) Flip-flop between lipid monolayers

to the other, the so-called flip-flop process, is, on the other hand, extremely slow, being of the order of hours, possibly days. In real biological membranes, special membrane proteins, so-called flippases, facilitate the redistribution of lipid molecules between the two monolayer leaflets.

The actual values of the rates of the different dynamical processes depend on the type of lipid molecule in question. Furthermore, there is some temperature dependence as well as a significant dependence on the state of matter of the lipid bilayer. If the lipid membrane is taken into a solid phase, all dynamical processes slow down significantly. For example, lateral diffusion is slowed down at least a hundred times. This is probably the single most important reason why membranes stop functioning when taken into solid phases.

The diffusion of lipid and protein molecules in membranes can be monitored by a number of experimental techniques. The motion of single molecules can be detected by either single-particle tracking or by ultra-sensitive single-molecule fluorescence microscopy or fluorescence correlation spectroscopy. In single-particle tracking, a colloidal particle of a typical diameter of 40 nm is linked to the lipid or protein molecule and the particle's motion is then followed by computer-enhanced video microscopy. Figure 8.5 shows the trace of a fluorescently labeled lipid molecule that diffuses in a lipid bilayer. The spatial resolution of this kind of experiment is about 50 nm, and the time resolution is in the range of about 5 ms.

In addition to the dynamical modes of the individual lipid molecules, collective motion of different kinds involving many lipid molecules take place over a wide range of time scales. These motions include bilayer undulations,

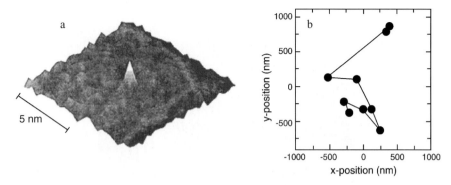

Fig. 8.5. (a) Fluorescence image of a single fluorescence-labelled lipid molecule in a POPE-POPC phospholipid bilayer. The peak in intensity signals a single molecule in the plane of the membrane. (b) Recording of a part of a diffusion trace of a single lipid molecule

bilayer thickness fluctuations, as well as collective diffusion of clusters of molecules within the plane of the membrane.

For comparison, integral membrane proteins are less lively. Proteins undergo more slow and restricted internal conformational transitions. Unless they are attached and anchored to the cytoskeleton, the proteins also diffuse laterally in the lipid bilayer. Their diffusion rate is typically a hundred times slower than lipid diffusion. If unrestricted, they would typically need about half an hour to travel over the range of a cell surface. Similarly, because of their larger circumference, they typically rotate around their axis about one full rotation during the time it takes a protein to travel a length corresponding to ten times its size.

The diffusion of lipids and proteins in biological membranes is often hindered because the motion takes place under certain restrictions. As mentioned, proteins can be attached to the cytoskeleton. Moreover, the membrane can be compartmentalized into various domains, which implies that proteins and lipids diffuse in an environment with obstacles, like a ship that navigates in an archipelago. In Sect. 11.3, we shall see how one can learn about lateral membrane structure by tracking the diffusional motion of individual molecules.

The many different dynamical processes that occur in lipid membranes are the reason why a lipid membrane in its liquid state is a very lively place indeed. The presence of lively dynamics in lipid membranes underscores the difficulty in working with a single unified model of biological membranes. Depending on the length and time scales one is interested in, a useful cartoon of a membrane will have different appearances. Since Fig. 8.1a is a snapshot in time and Fig. 8.1b is an average over time, these pictures do not provide the full information about possible dynamical aspects of trans-bilayer structure

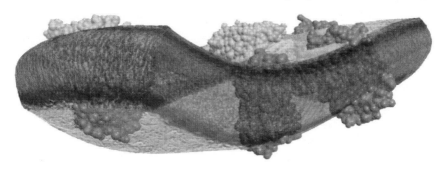

Fig. 8.6. Model of the lipid bilayer component of a cell membrane incorporated with integral membrane proteins (bacteriorhodopsin). The picture is drawn to scale, and it reflects averaging over fast dynamical modes. A 20 nm × 20 nm slap of a 5 nm thick lipid bilayer is shown. The time scale of view is in the range of 10^{-3} to 10^{-6} s. On this scale most molecular processes will appear blurred but not totally indiscernible. The transmembrane proteins are modelled by use of the X-ray coordinates for bacteriorhodopsin. Consistent with the slow time scale characterizing this picture, the protein surfaces have been slightly blurred

that may be relevant, for example, to how proteins and drugs are transported across membranes.

A different membrane model that appreciates the difficulty in presenting a liquid object with all this lively dynamics is presented in Fig. 8.6. This figure highlights the lipid bilayer component and details of the molecular structure of integral membrane proteins (bacteriorhodopsin). The picture is drawn to scale, and it reflects averaging over fast dynamical modes. The time scale of view is in the range of 10^{-3} to 10^{-6} s. On this scale most molecular processes will appear blurred but not totally indiscernible. For example, the very rapidly moving chains seen on the edges of the lipid bilayer are indicated by subtle texturing parallel to the chain axis. The texture reflects the order of the lipid chains, but the fatty-acid chains themselves are not seen. The membrane edge shading is based on information obtained from X-ray and neutron scattering. The shading used on the head group surfaces suggests the presence of small lipid domains. The picture shows clearly that the lipid bilayer displays large-scale bending fluctuations.

9 The More We Are Together

9.1 Phase Transitions Between Order and Disorder

Some of the most fascinating and spectacular events in nature arise when the matter changes state. Soft matter like lipid bilayers and membranes have their share of these phenomena. Let us, however, start with a simple and well-known example: water. Water in the form of ice melts upon heating into liquid water, which upon further heating turns into vapor. In this case, the matter water has three states, or so-called phases: a solid (ice), a liquid (water), and a gas (water vapor). Although all states are made of the same type of simple H_2O molecules, the three phases appear to the naked eye very different. They also turn out to have very different materials properties upon closer investigation.

The two transitions connecting the phases, i.e., melting and boiling, are called *phase transitions*. The phase transitions are in this case induced by temperature, they are so-called *thermotropic phase transitions*, and they occur at well-defined temperatures, the melting temperature (0°C) and the boiling temperature (100°C). Obviously, it is possible to go through the transitions in the reverse direction by cooling from the vapor phase. It is well-known that the boiling point of water depends on pressure. Boiling water for a cup of tea in the Himalayas only requires heating up to around 80°C. Similarly, it is well-known that adding something to the water will change its melting point. Adding salt to ice will reduce the melting point.

The description of the phase transitions for water above is rather general and applies basically to any kind of matter. Butter and fat are known to melt upon heating, alcohol evaporates when heated, and olive oil goes solid when frozen. A number of phase transition phenomena are less well-known, e.g., a magnet can lose its magnetization when heated, an insulator can become a conductor upon cooling (possibly even a super-conductor), a liquid crystal display can change color when heated, or a biological membrane can become solid and stop functioning when cooled.

Phase transitions are also called cooperative or collective phenomena and they are highly nontrivial consequences of the fact that many molecules interact with each other. The molecules act in a sort of social manner, and the more they are together the more dramatic are the transitions between the different states of matter. By being many together, the assembly of interacting molecules assumes properties that no single molecule possesses itself. For example, many water molecules together can form liquid water, although the particular properties associated with a liquid, such as fluidity and density, are not properties of a single water molecule.

The different phases of a material reflect different degrees of order. A solid is very ordered, typical of a crystal, where the molecules are arranged in a regular fashion, as illustrated in Fig. 9.1a. A liquid is more disordered, as shown in Fig. 9.1c, and the molecules in the liquid, although sticking together, diffuse around among themselves. Finally, the molecules in the gas in Fig. 9.1d hardly feel each other and the phase is very disordered. The three phases are distinguished by different densities. A special possibility exists of a solid, a so-called amorphous solid, as shown in Fig. 9.1b. The amorphous solid, which under some conditions is called a glass, has almost the same density as the crystalline solid but the molecules are positioned irregularly and often display very low mobility. This phase is similar to a very viscous liquid of the same density. Hence, this phase is in some sense both a liquid and a solid. As we shall in see in Sect. 9.4, the task carried out by cholesterol in lipid membranes is to produce a special phase of the type in Fig. 9.1b, which is in between order and disorder and where the molecules furthermore have the freedom to move around. In order to see this, we have to introduce an extra, necessary complication concerning the properties of the molecules.

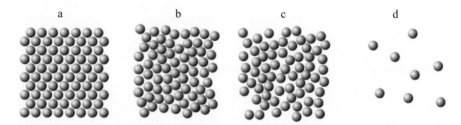

Fig. 9.1a–d. Two-dimensional representation of solid and liquid phases of matter composed of molecules that are spherically symmetric and have no shape. (**a**) Crystalline solid, (**b**) amorphous solid or very viscous liquid, (**c**) ordinary liquid, and (**d**) gas

It was tacitly assumed in the description of the phases in Fig. 9.1 that the molecules are isotropic and spherical, i.e., they have no internal degrees of freedom. We now proceed to the next level of complication where the molecules have a nonspherical shape, e.g., prolate and shaped like a cigar. Whereas ordering of the spherical molecules can only involve their positions in space, i.e., the positional (translational) degrees of freedom, prolate molecules have an extra possibility of displaying order and disorder via the direction of their long axis. Order would here imply that the molecules tend to orient their long axes in the same direction. We refer to the direction of the molecules as

9.1 Phase Transitions Between Order and Disorder

an internal degree of freedom, e.g., an orientational or configurational degree of freedom.

A number of new phases, so-called *meso-phases*, in between those of solids and liquids, can now be imagined. Some examples are given in Fig. 9.2. In the liquid phase in Fig. 9.2d, the molecules are disordered with respect to both the translational degrees of freedom and the orientational degree of freedom. In contrast, in the solid (crystalline) phase, order prevails in both sets of degrees of freedom. New phases can arise in between. The meso-phases have elements of order as well as disorder. In Fig. 9.2c, the positions of the molecules are disordered as in a liquid, but their long axes have a preferred direction, i.e., the collection of molecules are ordered orientationally. In Fig. 9.2b, the positions of the molecules have some element of order by being localized in a set of parallel planes that have a fixed distance as in a crystal. Within each of these planes, the positions are however disordered as in a two-dimensional liquid. The molecules are additionally ordered orientationally. Finally, in Fig. 9.2a the positions of the molecules within each plane are also ordered as in a crystal. The two meso-phases are called *liquid crystals*, respectively, a smectic (b) liquid crystal and a nematic (c) liquid crystal. A lipid bilayer is an example of a liquid crystal of the smectic type.

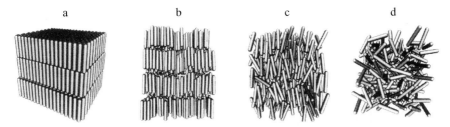

Fig. 9.2. Prolate molecules exhibiting solid crystalline (**a**) and liquid (**d**) phases and two intermediate liquid-crystalline meso-phases: smectic (**b**) and nematic (**c**)

Just as for the simpler systems illustrated in Fig. 9.1, the different phases in Fig. 9.2 are connected by phase transitions. The phase transitions can be triggered by changing temperature, i.e., they are thermotropic phase transitions. Increasing temperature will lead to transitions going from left to right in Fig. 9.2.

There is a large number of different liquid crystalline phases possible. More than 10% of all organic chemical compounds known today display one or another type of liquid crystalline phase. Liquid crystals are widespread throughout nature. Living matter is no exception. Being smectic liquid crystals, lipid membranes can therefore be considered nature's preferred liquid crystals. Hence, living matter is based on liquid-crystal technology.

Phase transitions are conventionally divided into two types: discontinuous transitions (first-order transitions) and continuous transitions (second-order transitions or critical-point phenomena). At a discontinuous transition, the degree of order in the system changes discontinuously at the transition temperature, whereas it changes in a continuous manner at a continuous transition. Furthermore, at a continuous transition, strong fluctuations prevail. A discontinuous transition can often be driven into a continuous transition by varying a suitable parameter, e.g., pressure. As an example, the boiling of water at 100°C and 1 atm is a first-order transition, and there is a large discontinuous jump in the density (degree of order) going from liquid water to vapor. By increasing the pressure, this jump can be gradually diminished and brought to vanish at 218 atm. At this so-called critical pressure, water evaporates according to a continuous transition at a corresponding critical temperature of 374°C, where there is no density difference between liquid water and vapor. The phase transitions of most thermotropic liquid crystals are of first order, whereas the magnetization of an iron magnet displays a critical-point phenomenon and vanishes continuously at the so-called Curie temperature.

9.2 Lipids Have Phase Transitions

As described in Chaps. 3 and 4 and as illustrated in Figs. 3.4 and 4.4, lipids in water form a number of different supramolecular aggregates. Lipid aggregates in water can be considered phases and states of matter. The phases are induced by varying the water concentration, and phase transitions between the different aggregate forms can in many cases be triggered by changing the water concentration under isothermal conditions. Such transitions are called *lyotropic transitions*.

For example, some lipid lamellar phases can undergo a transition to an H_I phase by increasing the water content and to an H_{II} phase by decreasing the water content. Alternatively, such lyotropic transitions can be triggered by changes in composition, by varying physico-chemical conditions, or by biochemical input, exploiting the fact that lipids speak the language of shape described in Chap. 4. The example described in Sect. 4.4 involving the microorganism *Acholeplasma laidlawii*, which maintains homeostatic control by changing the lipid composition of its membrane, is an excellent illustration of biological regulation by moving around with a lyotropic phase transition from a lamellar phase to an inverted hexagonal phase. *Acholeplasma laidlawii* arranges for its membrane to have a composition that positions the lamellar-hexagonal phase transition temperature about ten degrees above the ambient temperature.

In addition to phase transitions between phases of different morphology, lipid aggregates undergo a number of internal phase transitions without changing morphology. We shall here concentrate on lipid phase transitions within lamellar symmetries, specifically, in lipid monolayers as considered in Chap. 10 and in lipid bilayers as dealt with throughout the rest of this book, and, in particular, in Chaps. 11 and 12. Certain large-scale morphological transitions involving shape changes of large liposomes and whole cells, exemplified by red blood cells, were discussed in Chap. 6.

Aqueous dispersions of phospholipid bilayers in the form of uni-lamellar or multi-lamellar vesicles, as illustrated in Fig. 3.4, display a series of thermotropic phase transitions. This is exemplified in Fig. 9.3, which shows the specific heat as a function of temperature for a dispersion of multi-lamellar bilayers of DPPC. The specific heat is a measure of the heat capacity of the system, i.e., how much heat has to be supplied to raise the temperature one degree. The specific heat has in this case two peaks. A peak in the specific heat is an indication of a phase transition. The heat contained in the peak is a measure of the heat of transition, that is the amount of heat that has to be supplied in order to facilitate the transition. The two peaks separate three phases with trans-bilayer structures, which are illustrated schematically in Fig. 9.3. The presence of these transitions is general for all PC lipids. Other lipids with different fatty-acid chains and different head groups need not have all these transitions. However, the one appearing at the highest temperature in Fig. 9.3 is generally found for all phospholipids. This transition is called the *main phase transition*. The main phase transition is the intra-bilayer phase

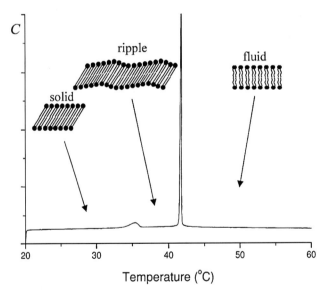

Fig. 9.3. Phase transitions and phases in DPPC lipid bilayers. Specific heat, C, as a function of temperature. The inserts show schematic illustrations of the trans-bilayer structure in the different phases separated by the phase transitions that are signalled by peaks in the specific heat. The small peak separating the solid (solid-ordered) and the ripple phase corresponds to the pre-transition, and the large peak separating the ripple phase and the fluid (liquid-disordered) phase corresponds to the main transition

transition that is believed to be of most importance for membrane biology. We shall use PC lipids to illustrate the nature of the main transition and how its properties are reflected in membrane function. The main transition is a first-order transition, although it is often found to be associated with rather strong fluctuations.

The main transition is characterized by a transition temperature, T_m, where the specific heat attains its maximum, and a heat (enthalpy) of transition, ΔH, which is a measure of the amount of heat that has to be supplied to the system for the transition to take place. T_m and ΔH are larger the longer the fatty-acid chains are. For increasing degree of unsaturation, the transition occurs at progressively lower temperatures.

The heat of transition corresponds to a transition entropy

$$\Delta S = \Delta H / T_m . \tag{9.1}$$

It turns out that this transition entropy is very large, around $15\,k_B$ per molecule for DPPC, where k_B is Boltzmann's constant. It is instructive to insert this number into Boltzmann's formula,

$$\Delta S = k_B \ln \Omega , \tag{9.2}$$

which gives a rough estimate of how many micro-states, Ω, of the system (per molecule) are involved in the transition. The resulting number is $\Omega \sim 10^5\text{--}10^6$. This is a very big number. The only available source for all these states is the conformations of the two long, flexible fatty-acid chains of the lipid molecule. An illustration of the richness of these states was given in Fig. 4.3.

We therefore conclude that the main transition is associated with a melting of the lipid molecules in the sense that the molecules below the transition have fairly ordered chains, whereas above the transition the chains are more disordered. This picture of the transition has been confirmed by a variety of spectroscopic measurements. This kind of structural change was indicated in Fig. 9.3 and is illustrated in greater detail in Fig. 9.4. Further structural and rheological studies have shown that the phase below the transition not only has ordered fatty-acid chains; the lipid molecules are at the same time arranged in a regular structure as in a crystalline solid. In contrast, the lipid molecules in the phase above the transition are positionally disordered as in a liquid and subject to rapid lateral diffusion.

Therefore, a complete description of the two phases calls for two labels in the same way as we saw in relation to liquid crystals in Sect. 9.1: one label (ordered, disordered) that refers to the conformational (internal) degree of freedom of the fatty-acid chains and another label (solid, liquid) that refers to the positional degree of freedom. We shall consequently call the phase below the transition the *solid-ordered* phase and the phase above the transition the *liquid-disordered* phase. This labelling shall turn out to be very helpful for describing the effect of cholesterol on the lipid bilayer phase transition, as

9.2 Lipids Have Phase Transitions

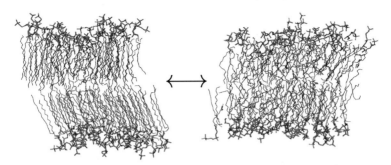

Fig. 9.4. A slab of a lipid bilayer illustrating the structural changes during the main phase transition. The picture is obtained from a Molecular Dynamics simulation on a DPPC bilayer in water using an atomistic model. To the *left* is shown the solid-ordered phase and to the *right* the liquid-disordered phase

described in Sect. 9.4. As we shall see, cholesterol is capable of inducing a new membrane phase which is in between the two: the *liquid-ordered* phase.

In order to illustrate the degree of fluidity and disorder in the liquid-disordered phase, Fig. 9.5 shows a snapshot from a computer simulation calculation on a model using a full-scale atomistic description.

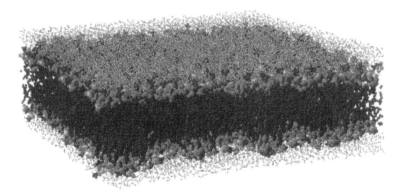

Fig. 9.5. A slab of a lipid bilayer illustrating the liquid-disordered phase. The picture is obtained from a Molecular Dynamics simulation on a DPPC bilayer in water using an atomistic model

The illustration of the two bilayer phases in Fig. 9.4 suggests that the transition in the bilayer is accompanied by a considerable decrease in bilayer thickness, Δd_L, and at the same time a substantial area expansion, ΔA, which is typically as large as 10–15%. It turns out that the volume per lipid is only changing a few percent during the transition. Hence, the thickness change and the area change are reciprocally related, i.e., $\Delta A \Delta d_\mathrm{L} = $ a constant. In

Sect. 8.3, we discussed how the different hydrophobic and hydrophilic parts of the bilayer contribute to its total thickness.

Figure 9.3 suggests that the solid-ordered phase displays an additional modulated feature over an intermediate range of temperatures. This is the so-called ripple phase, which is characteristic for PC lipids. The ripple structure is pictured in more detail in Fig. 9.6, which shows the surface of a DPPC lipid bilayer in water imaged by atomic force microscopy. The transition between the solid phase and the ripple phase is termed the *pre-transition*.

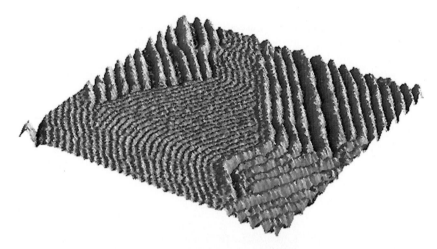

Fig. 9.6. Atomic force microscopy picture of the surface of a DPPC lipid bilayer in water that forms a ripple phase below the main transition. There are two types of ripples with periodicities of 13 nm and 26 nm, respectively. The picture is 600 nm × 600 nm

9.3 Mixing Different Lipids

When more than one type of lipid molecule is present in a bilayer, a more complex behavior results. There is no longer a single transition described by a single transition temperature. The transition now takes place over a range of temperatures where the system separates in more than one phase. One speaks about phase equilibria and phase separation. The social and cooperative behavior of the lipids has developed into a sort of separatism, where lipid molecules over a certain range of temperatures and compositions have a preference for separating out together with other molecules of their own kind.

The underlying physical mechanism for phase separation is simple: Lipid molecules of the same kind often have a stronger attractive interaction than lipids of different kinds. The resulting phase behavior is conveniently described in a so-called phase diagram, as illustrated for three binary lipid mixtures in Fig. 9.7. A phase diagram of the type shown indicates the actual phase of the mixed system for varying composition and temperature. The three mixtures in Fig. 9.7 were chosen in order to show how an increasing difference between the lipids that are mixed leads to an increasingly complex phase behavior. In the present case, the difference between a pair of lipids is simply the difference in the lengths of their fatty-acid chains. Lipid molecules that are more closely matched in chain length have a larger preference for being together. We saw in Fig. 9.3 that a peak in the specific heat arises at the main transition. In Fig. 9.8, the specific heat is shown for mixtures of DMPC and DSPC. For each composition, the specific heat is seen to have two peaks that occur at the boundaries in the phase diagram in Fig. 9.7.

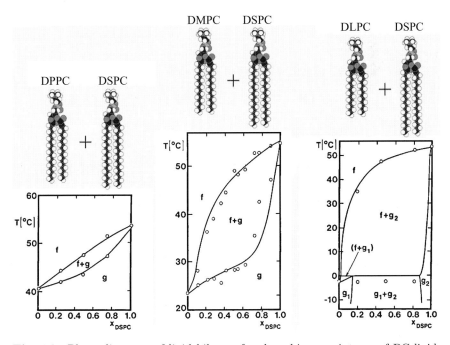

Fig. 9.7. Phase diagrams of lipid bilayers for three binary mixtures of PC lipids with different fatty-acid chain lengths. f denotes the liquid-disordered phase and g denotes solid-ordered phases

The DMPC-DPPC mixture has the weakest tendency for phase separation of the three mixtures considered in Fig. 9.7. In the phase separation region, the bilayer splits into two phases with different composition. The lipid species with the lowest value of the phase transition temperature, in this case DMPC, is enriched in the liquid-disordered phase, whereas the lipid with the highest

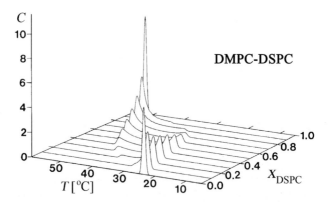

Fig. 9.8. Specific heat, C, as a function of composition for mixtures of DMPC and DSPC in a lipid bilayer

transition temperature, i.e., DPPC, is enriched in the solid-ordered phase. The phase-separation behavior is getting more pronounced for the DMPC-DSPC mixture that has a larger incompatibility in chain lengths. Finally, the DLPC-DSPC mixture develops an added complication in that the solid-ordered phases of the two lipids do not mix at low temperatures. Hence, a region of solid-solid phase separation arises.

Phase separation is in principle a macroscopic phenomenon, i.e., the separating phases are large. We shall in Chap. 11 remark that the phase separation is not necessarily fully developed in these lipid mixtures. Instead, the system is found to be organized in a large number of small lipid domains. As we shall discuss in Sect. 11.4, these domains are a mode of organizing membranes on a small scale, which may be important for many aspects of membrane function, e.g., binding of proteins, fusion, permeability, and enzyme activity.

Lipid mixtures can under appropriate conditions of temperature and composition develop critical-point phenomena, as discussed in Sect. 9.1 above. These conditions are referred to as critical mixing. Below in Sect. 9.4 we shall show that cholesterol when mixed into a phospholipid bilayer can lead to such a situation. At critical mixing, the mixture is subject to strong fluctuations in local composition, and large domains enriched in one of the mixed lipid species are transiently formed, as also discussed in Sect. 11.2.

9.4 Cholesterol Brings Lipids to Order

Being an amphiphilic molecule (cf. Fig. 2.5a), cholesterol easily incorporates into lipid bilayers with its hydrophilic –OH head group at the bilayer-water interface and the steroid skeleton inside the hydrophobic core. The cholesterol molecule barely spans one monolayer leaflet of a typical bilayer, as illustrated in Fig. 9.9. Considering the different types of ordering that pertain to the lipid

9.4 Cholesterol Brings Lipids to Order

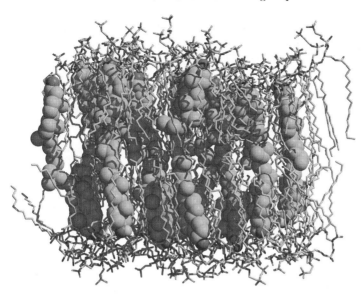

Fig. 9.9. Patch of a fluid DPPC lipid bilayer incorporated with 20% cholesterol. The picture is obtained from Molecular Dynamics simulations. The DPPC molecules are shown in thin lines and the cholesterol molecules are highlighted in a space-filling representation

molecules in a bilayer, cf. Fig. 9.4, it appears that cholesterol has a problem when coming to terms with the life in a dense lipid bilayer.

On the one side, cholesterol, due to its hydrophobically smooth and stiff steroid ring structure, has a preference for having conformationally ordered lipid chains next to it since they provide for the tightest interactions. From this perspective, cholesterol prefers the solid-ordered lipid phase. On the other side, the solid-ordered phase is a crystalline phase with dense packing order among the lipid molecules. The cholesterol molecule, with its own peculiar size and shape, does not fit well into this packing order, whereas there is plenty of free space to squeeze into in the liquid-disordered phase. From this perspective, cholesterol prefers the liquid-disordered phase. Hence, the cholesterol molecule becomes frustrated when presented with the two different lipid phases. This frustration and the way cholesterol finds a way of releasing the frustration hold the key to understanding not only the effect of cholesterol on the physical properties of lipid membranes, but also the role of sterols in the evolution of higher organisms and their membranes. We shall return to sterols in the context of evolution in Chap. 14.

Cholesterol releases the frustration by introducing a new phase, the *liquid-ordered* phase, first proposed in 1987 by the Danish biophysicist John Hjort Ipsen. The liquid-ordered phase is in between the two normal lipid bilayer phases. The resulting phase diagram is shown in Fig. 9.10. This diagram

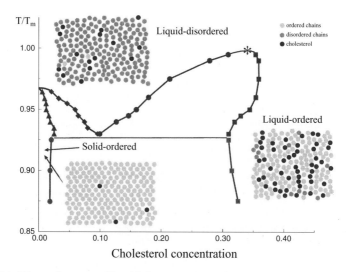

Fig. 9.10. Phase diagram of lipid bilayers with cholesterol. A simple representation is given of the lateral structure and organization of the bilayers as composed of ordered and disordered lipid chains and cholesterol. The critical point is marked by an asterisk. The sterol concentration is given in mole%

shows that cholesterol stabilizes the liquid-ordered phase over a wide range of compositions and temperatures. A particular feature of the diagram is that it appears to close on the top in a critical mixing point beyond which the liquid-ordered and the liquid-disordered phases become indistinguishable. Around this critical point it is expected that dramatic density and compositional fluctuations will occur. These fluctuations may be a source for the small-scale structures found in membranes containing large amounts of cholesterol, cf. Sect. 11.3.

This liquid-ordered phase is a genuine liquid with positional disorder and high lateral mobility of the membrane molecules. Furthermore, the lipid chains have a substantial degree of conformational order. When introduced into the liquid membrane phase, cholesterol leads to a large increase in membrane thickness. The thickening provides for larger mechanical coherence and less flexible bilayers. It also makes the bilayer tighter, as described in Sect. 12.1.

Consequently, cholesterol has a remarkable dual effect on membranes. It makes the membranes stiffer, but retains the fluidity required for membrane function. In a way, cholesterol acts as an anti-freeze agent. No other molecule is known to have a similar dramatic effect on lipid membrane behavior except the other higher sterols like ergosterol. Based on this insight we can qualify our discussion of what fluidity of membranes actually means.

The liquid membrane phases, whether it be liquid-ordered or liquid-disordered, are the membrane states that should be associated with membrane fluidity that is so central to the celebrated Nicolson-Singer model described in Sect. 7.2. However, fluidity is not a well-defined physical property. At best it is a loose term that covers the lively dynamics of the liquid phases of membranes. The trouble is that a lipid bilayer can be "fluid" in more than one sense of the word. As we have seen, the description of lipids requires at least two fundamental sets of degrees of freedom, the positional (translational) degrees of freedom and the internal (conformational) degrees of freedom. A lipid bilayer can exhibit fluidity in both sets of degrees of freedom. Hence, if fluidity is meant to imply fast diffusion, it refers to dynamic disorder in the translational variables. However, if fluidity is meant to reflect the fact that the fatty-acid chains can be conformationally disordered or melted, it refers to disorder in the internal degrees of freedom of the chains. A lipid bilayer can exhibit fluidity in both sets of degrees of freedom, as in the liquid-disordered phase, or in one of them, as in the liquid-ordered phase that can be induced by the presence of cholesterol.

10 Lipids in Flatland

10.1 Gases, Liquids, and Solids in Two Dimensions

We are used to studying processes and structures in nature as three-dimensional phenomena. Lipid bilayers and biological membranes also live in three-dimensional space. As pointed out and described in Chap. 6, membranes, due to their incredibly small thickness compared to their extension, are in many respects like two-dimensional systems imbedded in a three-dimensional space. Still, due to curvature, free membranes are not truly two-dimensional, and there is distinct coupling between the in-plane degrees of freedom and the curvature into the three-dimensional world. There are possibilities, however, of effectively fixing lipid membranes to stay in two spatial dimensions by confining them to a well controlled two-dimensional interface. This confinement can be provided by an air-water interface, where a lipid monolayer, as shown in Fig. 3.4a, is trapped due to the very limited solubility of the lipids in the water subphase. A lipid bilayer can also be caught in a flat configuration on a hydrophilic solid surface, as described in Sect. 11.2.

When spread on an air-water interface, water-insoluble amphiphilic molecules like long-chain fatty acids and phospholipids, form mono-molecular layers called *Langmuir films*. It was mentioned in Sect. 7.2 that the discovery of membranes being bimolecular layers was made using Langmuir film formation of the lipid contents of red blood cell membranes. The remarkable observation made by Langmuir almost a century ago was that monomolecular films can be compressed, using a simple film balance, to form two-dimensional versions of the well-known states of matter in three dimensions: gas, liquid, and solid. For the two dimensional phases, film area (A) and two-dimensional film pressure (Π) play the roles volume and hydrostatic pressure do in three dimensions. By using the film balance, as shown in Fig. 10.1, one can investigate the phase behavior and the phase transitions of the film simply by recording A as a function of Π for different temperatures. The film pressure is related to the interfacial tension, γ, as $\Pi = \gamma^* - \gamma$, where γ^* is the surface tension of pure water. An example of a resulting isotherm obtained from this procedure is given in Fig. 10.1. The two horizontal portions on the isotherm signal two phase transitions, one from a gas phase to a liquid phase and one from a liquid phase to a solid phase.

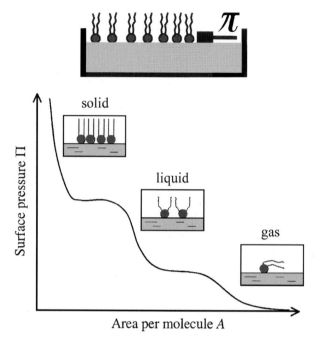

Fig. 10.1. Experiments on a monomolecular film (Langmuir film) of amphiphilic molecules in two dimensions. In the *top* is shown a cut through a Langmuir trough where the area, A, and the two-dimensional pressure, Π, of the film can be varied by moving the solid bar. In the bottom is given a schematic illustration of an isotherm that provides the relationship between A and Π. An indication of the structure of the gas, the liquid, and the solid phases is given

The peculiar characteristics of a lipid monolayer film in a solid phase can conveniently be investigated by assembling the lipid monolayer on a gas bubble in water, as shown in Fig. 10.2. The mechanical properties of the thin shell can be studied by micro-pipette aspiration techniques by which a pipette exerts a suction pressure on the bubble. The left-hand panel in Fig. 10.2 shows that the solid lipid shell on the bubble can be deformed by the pressure and that the deformed shape persists after the bubble is released again. Subsequently, sucking somewhere else on the deformed bubble restores the spherical shape. If the lipid monolayer were in the liquid phase, it would immediately assume a spherical shape in the non-pressurized state due to the interfacial tension. The solid lipid shell, like any other solid subjected to shear forces, builds up a strain in response to the tension exerted by the pipette suction pressure. The solid shell, furthermore, has crystalline grains, as illustrated by the fluorescent microscope image in the right-hand panel of Fig. 10.2.

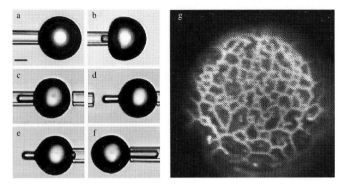

Fig. 10.2. Solid DSPC lipid shells on gas bubbles in water. The panel to the left shows how a suction pressure exerted by a micro-pipette in steps from a to f leads to a deformation of the solid shell which maintain its shape after release from the pipette until another pipette exerts a suction pressure on the other side of the bubble, leading to a restoring of the spherical shape. The diameter of the bubble is 15 μm. To the right in frame g is shown a fluorescence microscope image of the bubble exposing the crystalline grain structure where the grain boundaries are highlighted by a special fluorescent lipid dye that is incorporated into the monolayer

10.2 Langmuir and Langmuir-Blodgett Films

Lipids lend themselves readily to form Langmuir monolayers on air-water interfaces and to being investigated by the film-balance technique described above. The film is prepared by first dissolving the dry lipid material in some organic solvent. The solution is subsequently applied to the air-water interface by a syringe. After the organic solvent has evaporated, the lipid has formed a monomolecular layer at the water surface. A series of isotherms for DMPC monolayers are shown in Fig. 10.3. In this case, only the condensed solid and liquid phases of the film have been investigated.

At low temperatures there is a clear, almost horizontal portion of the isotherm signaling a transition from a liquid to a solid phase. This phase transition is analogous to the main phase transition in lipid bilayers described in Sect. 9.2. In the liquid monolayer phase, the fatty-acid chains are disordered and melted, whereas they are conformationally ordered in the solid monolayer phase. Hence, again in analogy with lipid bilayers, the liquid phase film is thinner than the solid phase film. A similar behavior is found for other lipid monolayer films, e.g., DPPC. It is interesting to remark that in lipid monolayer assays the transition is usually driven by pressure, and both pressure and temperature can be varied by the experimenter. This is in contrast to the main transition in bilayers that normally is induced by temperature, and the conjugate lateral two-dimensional pressure cannot easily be controlled in the experiment. It should be noted that this two-dimensional lateral pressure is different from the three-dimensional hydrostatic pressure that we discuss in Sect. 19.2.

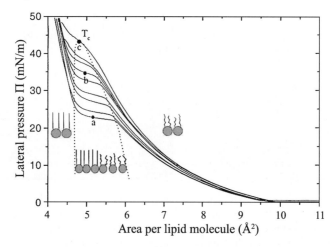

Fig. 10.3. Isotherms and phase diagram of a DMPC lipid monolayer at the air-water interface. The curves from a to c correspond to increasing temperature. The critical point at a temperature T_c is indicated. The region inside the *dotted* lines is the solid-liquid coexistence region, separating the solid phase to the *left* and the liquid phase to the *right*. Above the critical point, the two phases are identical

If the amount of lipid material spread on the interface is known, it is possible from the monolayer experiment to determine the average area per molecule. Furthermore, it is possible to monitor the effect of compounds that interact with lipids by introducing these compounds into the water subphase and studying the resulting change in area or lateral pressure. This is a widely used technique to investigate lipophilic drugs and the extent to which they bind to and penetrate lipid membranes.

Figure 10.3 shows that the flat portion of the isotherm for the DMPC monolayers is diminished as the temperature is increased. At a certain temperature, T_c, the isotherm passes smoothly from the liquid phase to the solid phase. This point is a critical point in the sense described in Sect. 9.1. Beyond the critical point, there is no longer a phase transition and the liquid and solid monolayer phases are identical. At the critical point, large fluctuations are expected. The questions are now how these fluctuations can be imaged directly and what the small-scale structure of the monolayer is in the range of the transition.

The determination of the molecular structure of condensed monolayer phases has turned out to be an elusive problem. Depending on the type of lipid material in question, a number of different solid phases of different crystalline structures have been revealed by X-ray crystallography. Another way to determine lateral structure is to use atomic force microscopy, which is a scanning probe technique that in principle can access scales down to the size of individual molecules and atoms, provided the sample is appropriately prepared. The atomic force microscope works by scanning a fine, sharp, needle-like tip positioned on a flexible cantilever, across the sample like a pickup on a record player. By this technique, the topographical landscape of the scanned surface can be obtained with a vertical and horizontal resolution that is better than a tenth of a nanometer. A high horizontal resolution however requires highly ordered and hard layers. If the sample is soft like a lipid monolayer in its liquid phase or in the transition region, the horizontal resolution will be much less, because the touch of the tip on the sample must be chosen to be very soft in order not to perturb the sample. Moreover, scanning techniques require that the sample is fixated.

Monolayers on air-water interfaces can be fixated by the so-called Langmuir-Blodgett deposition technique. As illustrated in Fig. 10.4, this technique involves transfer of the lipid monolayer to a solid substrate or support by pulling or pushing this substrate through the monolayer. Depending on whether the substrate is hydrophilic or hydrophobic, the lipids are deposited, respectively, with the head or the fatty-acid chain facing the support. By pulling the substrate through the monolayer film twice, a lipid bilayer can be formed. Special techniques exist to produce several layers. Such solid-supported bilayers are well controlled models of membranes and can be reconstituted with other membrane components such as proteins and sterols. The study of self-assembled solid films by Langmuir and Langmuir-Blodgett techniques has witnessed a renaissance in recent years because of its use to produce thin films of well-defined structure for use as devices in micro- and nano-electronics, as described in Sect. 20.1.

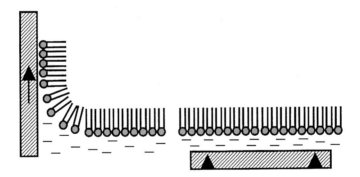

Fig. 10.4. Langmuir-Blodgett techniques can be used to transfer lipid monolayers to solid substrates to form supported monolayers or bilayers. Examples of vertical as well as horizontal transfer are illustrated

Since we are interested in lipid monolayers as models of biological membranes, we shall be concerned with the structure of lipid monolayer films predominantly in the liquid phase. We are now confronted with imaging a system that is a liquid or that possibly contains a mixture of solid and liquid domains, i.e., a very disordered and soft system. Therefore, it makes no sense to look for molecular details. We shall now see which techniques can reveal the lateral structure of liquid lipid monolayers.

10.3 Pattern Formation in Lipid Monolayers

The lateral structure and organization of lipid monolayer films on the air-water phase have been studied extensively by fluorescence microscopy and synchrotron X-ray scattering. Whereas scattering studies permit structural analysis of solid phases of the film at the molecular scale, fluorescence microscopy provides insight into the lateral organization of both liquid and solid phases on the micrometer and sub-micrometer scales. The trick used in fluorescence microscopy is to introduce a very small amount of probe molecules that often are lipid analogues. The probe molecules hence dissolve in the monolayer and can be seen in a fluorescence microscope because they contain a chemical group, typically a dye, that emits fluorescent light. If the fluorescent molecules are chosen such that they have a preference for a certain lipid phase, typically the liquid phase, the contrast in the microscopic picture can provide information about the extent of that phase and its lateral organization.

In Fig. 10.5 are shown examples of the type of lateral structures that have been observed by fluorescence microscopy of lipid monolayers at the air-water interface. In all cases, the solid phase and solid-phase domains are seen on the light background of the liquid phase that contains the fluorescent lipid molecules. Very complex patterns often arise as a consequence of the competition between the line tension of the lipid domains and the long-range electrostatic interaction between different domains.

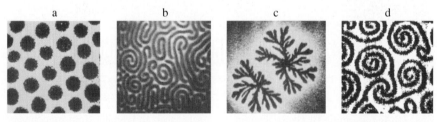

Fig. 10.5. Lateral structure of lipid monolayer films obtained by fluorescence microscopy. (**a**) Coexistence of liquid phase (*light*) and solid phase. (**b**) Striped pattern. (**c**) Fractal and dendritic solid patterns in a liquid-phase lipid monolayer after rapid compression. (**d**) Spiral solid domains in a lipid monolayer with cholesterol

The scale of the structures that can be observed by fluorescence microscopy is obviously limited by the wavelength of the light that is used in the microscope. Hence, structures substantially below the micrometer range

10.3 Pattern Formation in Lipid Monolayers

cannot be detected. In order to obtain a description of the lateral structure of lipid monolayers in the nanometer-micrometer range, other techniques have to be invoked. Atomic force microscopy is here an obvious choice. This technique has been applied to image the structure of various lipid bilayer and membrane systems in water and under physiological conditions. The approach is not without problems, since biological materials are soft and may easily be disturbed or damaged by the tip of the instrument.

In order to visualize the lateral structure of lipid monolayers by atomic force microscopy, the monolayer has to be fixated on a solid support, cf. Fig. 10.4, so it can be manipulated in the atomic force microscope. Under appropriate conditions, it is possible to transfer the monolayer film in a way that freezes the lateral structure in the monolayer onto the solid support without damage or significant distortion. In Fig. 10.6 are shown results for the lateral structure of a DMPC lipid monolayer as it is taken up toward its critical point, cf. the phase diagram in Fig. 10.3. A dramatic pattern formation is observed near the critical point. The patterns show lipid domains of different sizes, in the range of tens to hundreds of nanometers. The domains consist of liquid and solid-phase lipids in a fairly convoluted pattern. This is the signature of density fluctuations as they arise at a phase transition near a critical point.

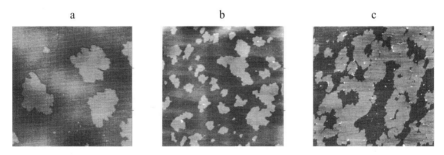

Fig. 10.6. Pattern formation observed by atomic force microscopy of lipid monolayers of DMPC approaching a critical point, corresponding to the points on the monolayer isotherms (**a–c**) in Fig. 10.3. The light regions represent lipid domains of the liquid monolayer phase in the (dark) solid monolayer phase. The two types of domains have different heights and can therefore be imaged in the microscope. The size of the images is $6\,\mu m \times 6\,\mu m$

Figure 10.7 provides a comparison of the pattern formation in lipid monolayers of DMPC and DPPC at their respective critical points. The critical fluctuations are seen to be more pronounced in the DMPC monolayer.

A similar type of pattern formation can be observed for mixtures of different lipids in monolayers treated as above. Figure 10.8a shows the pattern of a monolayer made of a 1:1 mixture of DMPC and DSPC. Two types of domains are observed, corresponding to segregated regions of the two

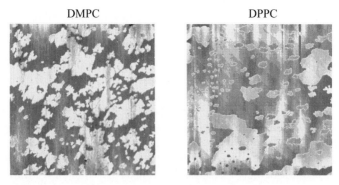

Fig. 10.7. Pattern formation observed by atomic force microscopy of lipid monolayers of DMPC and DPPC at their respective critical point. The size of the images is 25 μm × 25 μm

kinds of lipid molecules. Figure 10.8b shows a height contour plot across the domain pattern. The observed height difference corresponds to the expected difference in thickness of a DMPC and a DSPC lipid monolayer. This picture shows that although the lateral resolution of liquid films of lipids is limited to the range above tens of nanometers, the vertical scale is much better resolved. In fact, it is possible to obtain an estimate of the lipid monolayer thickness through this kind of experiment.

These results show that the cooperativity associated with the lipid monolayer phase transition provides a mechanism to structure the lipid monolayer on the nanometer scale. We shall in Sects. 11.1 and 11.2 see that the same mechanism is operative to forming lipid domains in lipid bilayer membranes via the main phase transition and the phase equilibria associated with that transition.

10.4 Lipids Make the Lung Work

Lipids and lipid monolayer films are critical for the functioning of the lung. Lipid polymorphism and lipid phase transitions, as discussed in Sects. 4.3 and 10.1, respectively, enter in a very direct way to both secure the lung from collapsing during exhalation and reduce the amount of work we have to do when we breathe.

The lung is a branched network of air-filled channels that terminate in a large number of small cavities called alveoli. A schematic illustration of an alveole is presented in Fig. 10.9. During inhalation the alveoli expand in order to take in oxygen, which subsequently is exchanged with carbon dioxide upon exhalation, when the alveoli again reduce their size. The total inner surface of the lung system is very large, about the area of a badminton court.

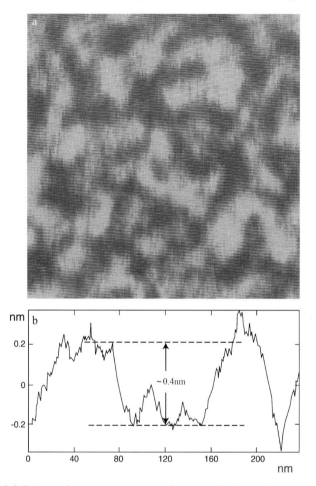

Fig. 10.8. (a) Pattern formation in a monolayer made of a 1:1 mixture of DMPC and DSPC observed by atomic force microscopy. The two types of domains correspond to segregated regions of the two kinds of lipid molecules. In (b) is shown a height contour across the domain pattern of the DMPC-DSPC mixture. The observed height difference of about 0.4 nm corresponds to the expected difference in thickness of a DMPC and a DSPC lipid monolayer

The mechanical functioning of the lung sounds simple but it actually presents a serious problem of stability due to the size of the alveoli. An alveole is small, typically with a radius, R, of a few hundred micrometers, which implies that it exhibits a high curvature R^{-1}. A curved surface is subject to a pressure difference across the surface, the so-called *Laplace pressure*. The Laplace pressure, p, is determined by

$$p = 2\gamma/R , \qquad (10.1)$$

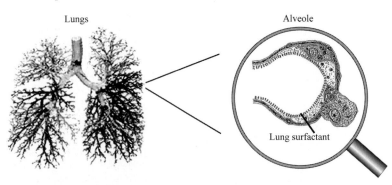

Fig. 10.9. Picture of the lungs illustrating the fractal structure of the airways and the alveoli. The enlargement to the *right* provides a schematic illustration of an alveole shown as an air-filled cavity surrounded by lung epithelium. On the surface of the epithelium toward the cavity is shown a lipid monolayer that contains mostly DPPC

where γ is the interfacial tension. This is the pressure that eventually makes soap bubbles burst.

During expiration, the increased air pressure in the small lung alveoli would tend to force more air out of the alveoli, which ultimately could lead to collapse of the lung. One way of solving this problem is to reduce the Laplace pressure by reducing the value of the interfacial tension γ. This is where lipids come in. Nature uses phospholipids as surfactants to spread as a mono-molecular film at the inner side of the lung alveoli in order to reduce the interfacial tension and hence the dangerous Laplace pressure. The mechanism of the active monolayer in lung function is not known in detail, and the full physiological importance of the lung surfactant layer is still an issue of debate. However, a gross picture has emerged. According to this picture, phospholipids and certain surfactant proteins, so-called pulmonary surfactants, are essential to making the lung work. Lack of lung surfactants, which could be induced by infectious diseases or a serious loss of lung surfactants during intensive care in a respirator, can lead to lung function failure. Interestingly, lung surfactants are formed only in the fetus' lungs in the late stages of pregnancy. Prematurely born infants can therefore suffer from the so-called respiratory distress syndrome and are in danger of dying. It is simply too energy demanding for them to inhale and expand the alveoli, and their lungs may collapse during exhalation. These babies need to be supplied with lung surfactant in order to survive.

The lung surfactant is by weight almost 90% phospholipid and about 10% protein. There are at least four different surface-active proteins involved. Most of the phospholipids are DPPC and about 5–10% are PG lipids. These lung surfactants lower the interfacial tension and reduce the Laplace pressure. Reducing the Laplace pressure of the alveoli is not enough, however, to make the

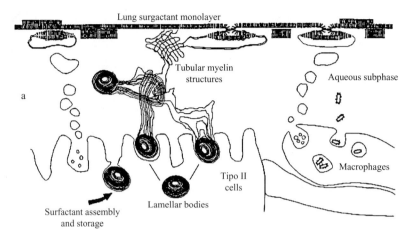
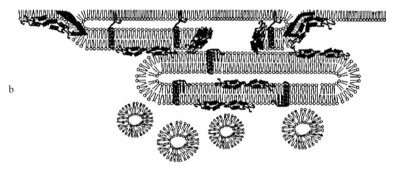

Fig. 10.10. Schematic illustrations of the lung surfactant monolayer at the surface of the lung alveoli. (**a**) The connected surfactant reservoir ranging from the cells that secrete the surfactants, across convoluted structures called lamellar bodies and tubular myelin structures, to the surface of the alveoli. (**b**) Possible fusion mechanism of the lung surfactant reservoir with the monolayer on the surface of the alveoli. Two types of membrane-bound proteins, SP-B and SP-C, are indicated to play a role in the folding and merging of the surfactant structures

lung work. The trouble is that the size of the alveoli varies dramatically during the lung cycle of inhalation and exhalation, the alveolus diameter changes typically from 250 μm to 160 μm. In order to maintain a low interfacial tension under these varying conditions, the lipid density at the interface has to change. This requires a reservoir of lipids with which the film can exchange material. Furthermore, in order to keep the interfacial tension sufficiently low, the active lipid monolayer has to be in the highly compressed solid phase.

It has been proposed that this reservoir is provided by lipid vesicles or lipids in a cubic or sponge phase, cf. Fig. 6.2d. This is illustrated in Fig. 10.10. The lipid reservoir is connected to the pathway of export of the lung surfactant from the special cells in the lung epithelium that secretes the

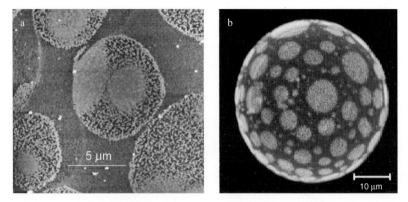

Fig. 10.11. Lipid domains in native pulmonary surfactant membranes consisting of lipids and proteins. (**a**) Atomic force microscopy image of a supported bilayer on mica. (**b**) Fluorescence microscopy image of a giant uni-lamellar liposome. The round domains in (**a**) are seen to have a granular structure. The individual spikes in this granular structure are presumably single lung surfactant proteins

surfactants. The reservoir can exchange lipids with the monolayer film by mechanisms that are not known in detail. A possible scenario is illustrated in Fig. 10.10, which also shows that the lung surfactant proteins are actively involved in the mechanics of the folding and fusion of the surfactant reservoir with the monolayer on the surface of the alveoli. The lipids in the lipid reservoir are likely to be in their liquid state. Upon incorporation into the monolayer, the lipids are believed to enter a solid phase. Upon exhalation, the excess lipids are squeezed out into the fluid state. These processes therefore require a lipid phase transition, as described in Sects. 9.2 and 10.1, to occur in the alveoli. Consequently, the lung surfactant layer has to have a composition such that its lipid phase transition is close to the physiological temperature. It is reassuring to note that the phase transition of the lipid, DPPC, which is the dominant lipid in lung surfactant, has its phase transition at 41°C, which is close to human body temperature. In fact, membranes made from natural lung surfactants have a phase transition very near 37°C.

It has recently been discovered that the layer of lung surfactants at physiological temperatures form a domain structure consisting of two kinds of lipid domains, as shown in Fig. 10.11. The coexisting domains are both fluid lipid phases, and it is seen that the lung surfactant proteins tend to accumulate in one type of the lipid domains. It has been surmised that cholesterol is an important regulator of this domain structure, which may have important consequences for the mechanical function of the lung.

11 Social Lipids

11.1 Lateral Membrane Structure

Being many molecules together in a bilayer membrane, lipids act socially, as described in Chap. 9, and organize laterally in the plane of the bilayer in a nonrandom and nonuniform fashion. In contrast to the trans-bilayer structure described in Chap. 8, the lateral bilayer structure and its molecular organization are less well characterized and their importance also generally less appreciated. This is particularly the case when it comes to the small-scale structure and micro-heterogeneity in the range of nanometers to micrometers. One reason for this is that this regime is experimentally difficult to access by direct methods. Another reason is that the small-scale structures are often dynamic and change over time. Finally, the strong influence on the thinking by the classical membrane models described in Sect. 7.2 has tended to make many researchers tacitly assume that the lipid bilayer is more or less a random mixture.

However, several physical mechanisms can lead to the formation of a highly nonrandom and nontrivial lateral organization of membranes. Firstly, proteins anchored to the cytoskeleton can provide effective fences or corrals that lead to transient or permanent membrane domains. Secondly, phase separation can occur, leading to large areas of different molecular composition. Finally, the molecular interactions between the membrane constituents (in particular, the lipids) lead to cooperative behavior and phase transitions, which can be associated with significant fluctuation effects, as described in Chap. 9. These fluctuations are the source for the formation of lipid membrane domains on different time and length scales. We shall focus on this type of domains here and return to domains involving proteins in Sect. 11.3.

Lipid bilayer fluctuations can be perceived as either local density variations or local variations in molecular composition. The range over which these variations occur is described by a so-called *coherence length*. The coherence length is a measure of the size of the lipid domains. Obviously, these domains need not be sharply defined and a certain gradual variation in the lipid bilayer properties is expected upon crossing a domain boundary. Lipid domains caused by fluctuations should be considered dynamic entities that come and go and that have lifetimes that depend on their size and the thermodynamic conditions. We refer to this type of domain formation as dynamic heterogeneity.

In order to develop some intuition regarding how dynamic heterogeneity could look like, we show in Fig. 11.1 some snapshot images of lateral bilayer structure obtained from computer simulation calculations on very simple models of one-component and two-component lipid bilayers. The only input to these simulations is the molecular interactions that act between nearest-neighbor lipid chains. Hence, the resulting dynamic lateral heterogeneity shown in Fig. 11.1 is a highly nontrivial consequence of the cooperative nature and social behavior of the lipid assembly. The snapshots demonstrate that one-component lipid bilayers near their phase transition develop a pattern of domains whose average size increases as the transition is approached from either side (Fig. 11.1a). In the solid-ordered phase, domains of lipids in the liquid-disordered phase arise, and vice versa in the liquid-disordered phase. The degree of heterogeneity is larger the shorter the lipid chains are (Fig. 11.1b). In the case of mixtures of lipids with different chain lengths, a pattern of density fluctuations develop in the liquid-disordered phase, and the domains are larger the larger the disparity in chain lengths (Fig. 11.1c). Finally, Fig. 11.1d shows a complex and convoluted nonequilibrium pattern of liquid-disordered and solid-ordered domains that coarsen as a function of time after a binary mixture has been cooled into the coexistence region. It is known from experiments that this coarsening process is exceedingly slow, with typical time scales on the order of hours. Hence, lipid bilayer mixtures may never reach complete phase separation on time scales of biological relevance.

Before we present the direct experimental evidence for lipid-domain formation in bilayers, we provide some indirect evidence that has been obtained from an experiment that was stimulated by the theoretical considerations above. This experiment builds on a spectroscopic principle of transfer of fluorescence energy between specifically labelled lipid molecules that are introduced as molecular probes in small amounts in the lipid bilayers in order to report back about their environment. If we use two types of labelled lipids, one type (the donor) that prefers one bilayer phase and another type (the acceptor) that prefers the other bilayer phase, and if two different probes upon close contact transfer fluorescence energy from the donor to the acceptor, then the measured fluorescence intensity would provide a measure of how many pairs of donors and acceptors are close in space. If the donors and acceptors each have their preferred phase or domain available in which to localize, there would be fewer contacts and the fluorescence intensity of the donor molecules would be higher than if there were only one type of phase or domain from which to choose. Hence, if lipid domains of the type shown in Fig. 11.1 exist, one would expect to see peaks in the donor intensity at phase transitions and at phase boundaries. This is indeed the case, as shown in Fig. 11.2. The figure shows a peak at the phase transition temperature

11.1 Lateral Membrane Structure 119

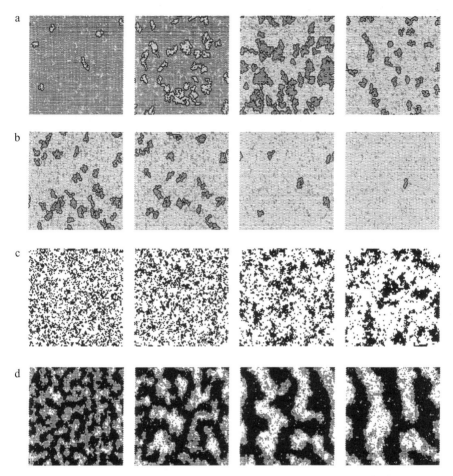

Fig. 11.1a–d. Snapshots of lateral bilayer structure obtained from computer simulation calculations on simple models. (**a**) A lipid bilayer of DPPC at temperatures around its phase transition. The temperature increases from *left* to *right*. The two left-most frames correspond to temperatures below T_m and the two right-most frames to temperatures above T_m. (**b**) Lipid bilayers of (from *left* to *right*) DMPC, DPPC, DSPC, and DAPC in the liquid-ordered phase at the same temperature, T/T_m, relative to their respective phase transition at T_m. (**c**) Lipid bilayers of binary lipid mixtures in the liquid-disordered phase. The difference in fatty-acid chain length (2, 4, 6, and 8 carbon atoms, respectively) increases from *left* to *right*. (**d**) The dynamic phase separation process as a function of time for a binary mixture of DMPC-DSPC that is brought suddenly from the liquid-disordered phase into the phase-coexistence region, cf. the phase diagram in Fig. 9.7. Time lapses from *left* to *right*

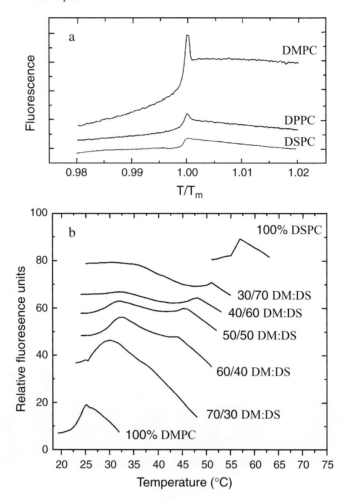

Fig. 11.2. (a) Fluorescence energy transfer data. Fluorescence intensity is shown as a function of temperature for lipid bilayers of DMPC, DPPC, and DSPC near their respective phase transition. (b) Fluorescence energy transfer data for binary lipid mixtures of DMPC-DSPC across the phase diagram, cf. Fig. 9.7

for the one-component lipid bilayers and that the peak is more intense and broader the shorter the lipid chain. Furthermore, it is found that there are peaks at the phase boundaries for the binary mixture, as well as a high level of intensity in the coexistence region. This set of data is strong indirect evidence of lipid domain formation and lateral heterogeneity in lipid bilayer membranes. Similar techniques have been used to detect domains in more complex membrane systems. Obviously, these fluorescence techniques cannot be used to determine an actual length scale of the lipid domains.

11.2 Imaging Lipid Domains

Direct imaging of the lateral structure and possible domain formation can be performed on individual bilayers using the same techniques applied to lipid monolayers in Sect. 10.3, specifically, fluorescence microscopy and atomic force microscopy. It requires, however, that the membranes be fixated in some way. Imaging by fluorescence microscopy exploits the possibility that different fluorescent probes with different colors can localize differently in different membrane phases and membrane domains. The contours of the domains then appear as contrasts between regions with different colors. These techniques have been widely used to image the surface structure of whole cells and fragments of real biological membranes.

Some very significant and definite evidence for the presence of lipid domains in well-defined model membranes has been obtained from fluorescence microscopy on giant uni-lamellar vesicles of diameter 50–100 µm. A range of different membranes have been investigated, including simple binary mixtures as well as lipid and protein extracts from real cell membranes. Figure 11.3 shows a gallery of images obtained for different giant vesicle membranes. The pictures show that on the length scales accessible by microscopy based on light, lipid domains occur in these membrane systems. The observed domains in the mixtures can be related to the different lipid phases. This relation can be established by a particular trick involving the use of special fluorescent probe molecules whose emitted light has a color that depends on the degree of lipid-chain order in the domain in which the probe resides. The high spatial resolution is obtained using so-called two-photon laser-scanning fluorescence techniques.

Fluorescence microscopy can also be applied to lipid bilayers supported on a solid hydrophilic surface of, e.g., glass or mica, as illustrated in Fig. 11.4. Such supported bilayers can be formed either by Langmuir-Blodgett deposition, as described in Sect. 10.2, or by fusion of lipid vesicles directly on the support. Single bilayers as well as multiple bilayers can be formed by these techniques. Once on the solid support, the lateral structure of these layers can be investigated by fluorescence microscopy or atomic force microscopy.

Figure 11.5a shows the image obtained for a supported lipid bilayer made of a DMPC-DSPC mixture using fluorescence microscopy. The solid DSPC domains in the fluid DMPC phase are clearly visible. In order to obtain a better resolution of the lateral structure on smaller scales, atomic force microscopy imaging can be applied. In Fig. 11.5b is shown an example of the visualization of the domain pattern in a 1:1 DMPC-DSPC lipid bilayer on a mica support. Since lipid bilayers are soft, it is often not possible using these techniques to faithfully image lateral domain formation, which only give rise to small or subtle variations in the membrane surface topography. However, tricks can be used to enhance the contrast by exploiting the height differences that can be induced by binding a ligand specifically to a receptor localized at the head groups of the lipids in a putative domain. An example

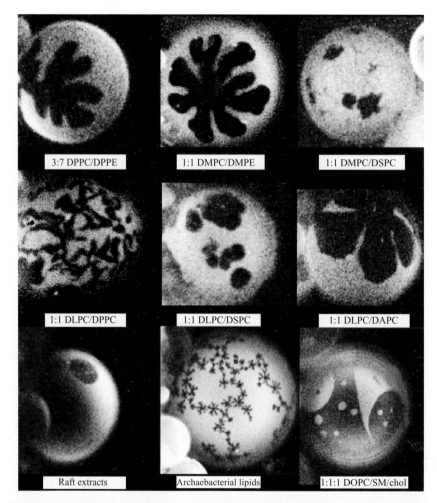

Fig. 11.3. Gallery of fluorescence microscopy images of the lateral structure of different lipid bilayers forming giant uni-lamellar vesicles. The typical size of the vesicles shown is 50 μm. The raft extracts refer to membrane compositions of lipids and proteins extracted from biological membranes with putative rafts (cf. Sect. 11.3)

of this method of developing domain structures is given in Fig. 11.5c, where a barely discernible domain pattern in the supported bilayer of a 1:1 DPPC-DAPC mixture becomes visible, as shown in Fig. 11.5d, upon adding avidin molecules that bind to a biotin moiety positioned at a small fraction of the DPPC lipid head groups. The avidin molecule has a diameter of about 5 nm, which provides for a height difference that can easily be picked up by the atomic force microscope.

11.2 Imaging Lipid Domains   123

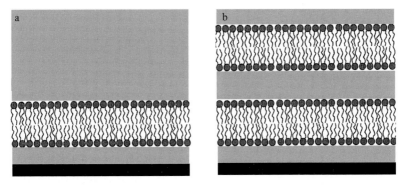

Fig. 11.4. Single (**a**) and double (**b**) lipid bilayers in water on a solid hydrophilic support, e.g., glass or mica

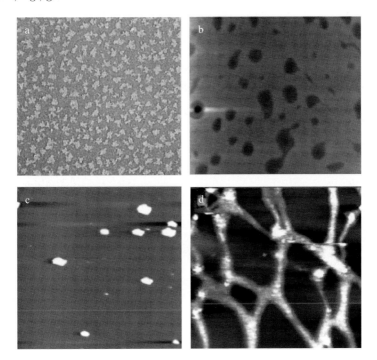

Fig. 11.5. (**a**) Fluorescence microscope image of a 1:1 binary mixture of DMPC and DSPC lipid bilayer on a mica support. The size of the image is 150 μm × 150 μm. (**b**) Atomic force microscopy images of a 1:1 DMPC-DSPC lipid bilayer on a mica support. The bilayer is in the solid-liquid phase separation region. The size of the image is 2 μm × 2 μm. (**c**) Atomic force microscopy image of a 1:1 DPPC-DAPC lipid bilayer on a mica support. The image has a domain structure that can barely be discerned. (The white spots are defects in the bilayer.) (**d**) The invisible domain structure in (**c**) is exposed by using a biotin-avidin amplification technique that enlarges the height differences between the domains and the background phase. The size of the images (**c**) and (**d**) is 6 μm × 6 μm

11.3 Lipid Rafting

The presence of small-scale lateral structure in biological membranes and its importance for biological activity have received increasing attention in recent years. The interest is fuelled by two types of information. Firstly, it was discovered by the single-particle tracking techniques described in Sect. 8.4 that labelled lipid or protein molecules performed a lateral diffusive motion, which suggested that they were temporarily confined to a small region of the membrane surface. An experimental recording of this behavior is given in Fig. 11.6, which shows the confinement and how the labelled particle after some transient time performs a jump into another confined zone where it continues its random diffusion.

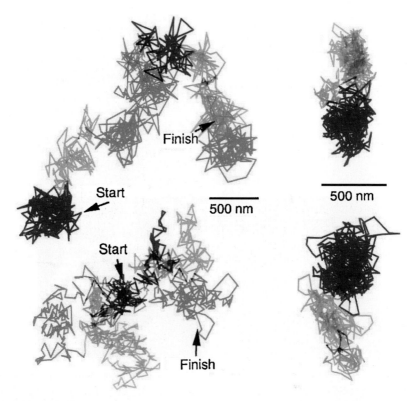

Fig. 11.6. Different traces of diffusion of proteins in membranes with an underlying domain structure. The four traces correspond to four different proteins

The other line of evidence derives from biochemical treatment of cold membrane samples dissolved in often rather harsh detergents such as Triton X-100. It was discovered that a certain fraction of the membranes was resistant to the detergent, and it was suggested that this fraction corresponds to supramolecular entities floating around in fluid membranes as a kind of *raft*. The rafts were surmised to behave as functional units supporting various functions to be discussed below in Sect. 11.4. So far, it remains unclear whether the proposed rafts were present before the addition of the detergent or were to some unknown extent induced by the preparation.

The accumulated evidence in favor of some kind of raft-like structures in a variety of membranes is now rather solid, although most of the evidence is indirect. A common characteristic of the rafts is that they contain high levels of cholesterol and sphingolipids, as well as saturated phospholipids. The presence of sphingolipids such as sphingomyelin or glycosphingolipids, which often have high phase transition temperatures, together with cholesterol, which promotes ordering of the lipid chain, led to the suggestion that the rafts have a structure similar to the liquid-ordered phase in the lipid-cholesterol phase diagram in Fig. 9.10. However, at present, there is only very limited direct evidence for the phospholipids in the rafts actually having a liquid-ordered structure.

A current pictorial and popular representation of a membrane raft is given in Fig. 11.7. This raft entity is imagined to be very small, possibly including only a few hundred molecules. Although illustrative, pictures like this are likely to be highly unrealistic since they neglect the dynamic nature of fluid membranes.

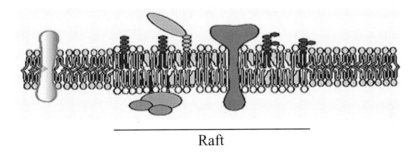

Fig. 11.7. Schematic illustration of a membrane raft consisting of a lipid patch enriched in sphingolipids, glycolipids, and cholesterol to which certain proteins are attached

Rafts are believed to be associated with peripheral as well as integral proteins that stabilize the rafts and function in connection with the rafts. Raft-like entities are, however, also found in simple lipid mixtures containing sphingomyelin, cholesterol, and saturated phospholipids. This is demonstrated in the fluorescence microscopy image of DOPC/SM/chol in Fig. 11.3. Furthermore, the fluorescence signal shows that the probe molecules experience a conformationally ordered lipid environment, suggesting the raft resembles a liquid-ordered phase. Obviously, the raft domains in Fig. 11.3 are very large and of the order of several micrometers. The question is whether there are smaller rafts in these simple raft mixtures that lack proteins. Some evidence for smaller rafts is provided in Fig. 11.8, which shows images obtained from atomic force microscopy of the same raft mixture now in the form of lipid bilayers on solid supports. Domains of micrometer and sub-micrometer size are seen in the image to the left, and the frame to the right demonstrates that these domains or rafts resist the treatment with detergent. Still, these domains in the model systems are much larger than the raft illustrated in Fig. 11.7. The lipid domains observed in the native pulmonary surfactant membranes shown in Fig. 10.11 may also be considered a kind of raft. In that case, it has been demonstrated that the formation of the domains is controlled solely by the lipids. Upon extraction of the pulmonary proteins, the domain pattern remains essentially unchanged.

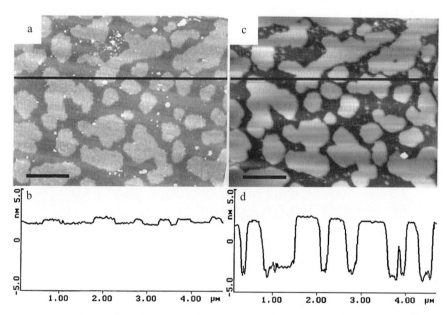

Fig. 11.8. Atomic force microscope images of lipid bilayers made of raft mixtures of 1:1 DOPC-sphingomyelin containing 25% cholesterol. Before (**a**) and after (**c**) treatment with the detergent Triton X-100 at 4°C. The images are 5 μm × 5 μm and the scale bars are 1 μm. Underneath the images are shown in (**b**) and (**d**) the corresponding cross sections of the height profile along the horizontal lines shown at the images above

11.4 Domains and Rafts Carry Function

The presence of domains and rafts in lipid bilayers and biological membranes highlights the need for a refined version of the model perception of membranes discussed in Sect. 7.3. These structured supramolecular entities may be seen as a way in which the membrane beats the nightmare of randomness and disorder in a liquid membrane, leading to the organization and ordering of the many different molecular species of the membrane. There is now accumulating evidence that domains and rafts also support aspects of membrane function.

Certain proteins seem to prefer association with rafts. Many of these proteins carry a hydrocarbon chain anchor that fits snugly into the tight packing of the raft. Recruitment of proteins to the rafts or detachment of proteins from the rafts can conveniently be facilitated by enzymatic cleavage or attachment of appropriate hydrocarbon chains. For example, long saturated fatty-acid chain anchors have an affinity for the ordered raft structure, whereas the more bulky isopranyl chain anchors prefer to be in the liquid-disordered phase outside the rafts.

Rafts have been shown to facilitate communication between the two monolayer leaflets of the bilayer and to be involved in cell surface adhesion and motility. Furthermore, there are indications that rafts are involved in cell surface signalling and the intracellular trafficking and sorting of lipids and proteins, as discussed in Sect. 15.1. It is interesting to note that some of these functions become impaired when cholesterol, which appears to be a necessary molecular requirement for raft formation, is extracted from the membranes.

A particularly important type of membrane domain is the so-called *caveolae*, which appear as invaginations of the membrane. Caveolae are specialized lipid domains enriched in cholesterol and glycosphingolipids, which are formed by the small transmembrane protein caveolin. Caveolae are found in a number of cell types, e.g., endothelial cells, and are involved in cholesterol transport, cytosis, and signal transduction.

The formation of lipid domains of a particular composition and structure implies differentiation and compartmentalization of the lipid bilayer that control the association and binding of peripheral (e.g., charged) macromolecules and enzymes. For example, water soluble, positively charged proteins (such as cytochrome c) exhibit enhanced binding to lipid membranes where a small fraction of negatively charged lipids form domains that have a local charge density large enough to bind the proteins. Conversely, the charged protein helps to stabilize the charged micro-domain. Enzymes like phospholipases and protein kinase C, as will be discussed in detail in Sects. 12.3 and 15.2 below, display variations in activity that correlate with the occurrence of small lipid domains.

The domain organization and the connectivity properties of the different membrane regions have consequences for the diffusional properties of membrane-bound molecules such as enzymes and receptors, and may, hence, control the kinetics and reaction yields of the associated chemical reactions.

Possible scenarios include accumulation and co-localization of receptors and ligands (such as drugs) in the same (small and specialized) membrane compartments via cooperative domain-organization processes and specific percolation events, as illustrated in Fig. 11.9.

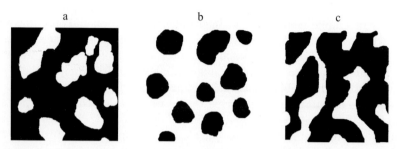

Fig. 11.9a–c. Schematic illustration of lateral domain organization of the surface of a membrane. (**a**) The white domains are isolated and molecules trapped in these domains cannot get in contact by diffusion. (**b**) The white region is connected and the molecules can get in contact by diffusion. (**c**) Bicontinuous situation where the membrane is split up into two convoluted and percolated structures

Lateral bilayer heterogeneity in terms of lipid domains furthermore implies changes in the macroscopic bilayer properties, e.g., lateral compressibility, bending rigidity, permeability, binding affinity for various solutes, as well as the way the bilayer mediates the interaction and organization of membrane proteins and peptides. This is the topic of the following chapter.

12 Lively Lipids Provide for Function

12.1 Leaky and Thirsty Membranes

It has been intensively discussed whether lipid phase transitions have any bearing on biological phenomena, and there are conflicting opinions concerning this question. The viewpoint advocated here is that lipid phase transitions as such are unlikely to be of direct relevance to most membrane functions. They are simply not robust enough to be of use in the delicate and specific regulation that is needed in biology. Having said this, we are left with the fact that lipids in lipid membrane assemblies are social; they behave in a cooperative manner that reflects the underlying phase equilibria. This has to be reflected in the way a membrane is organized and, ultimately, in the manner in which it functions. In order to give some hints as to which effects lipid phase behavior may have on membrane function, we present two simple examples on passive membrane functions that are affected by the fluctuations accompanying bilayer phase transitions.

The foremost mission of the lipid bilayer component of membranes is to act as a permeability barrier. Nonspecific passive permeation has to be avoided. The cell wants to use specific membrane-bound channels and pumps that work in a vectorial and controllable manner in order to keep track of the transport across membranes. The permeation of molecular species across lipid bilayers depends on both the diffusion rate and the solubility of the permeant in the membrane. The permeability, therefore, intimately reflects the inhomogeneous nature of the membrane, both transversely and laterally. Lipid bilayer permeability is our first example of how the lively lipids and their properties at the phase transition influence a passive membrane function.

Whereas lipid bilayers are moderately permeable to water, gaseous substances like CO_2 and O_2, small hydrophobic molecules like benzene, ions and larger molecular species such as glucose, amino acids, as well as peptides only pass very slowly across the bilayer. The passage of hydrophilic and charged compounds like ions is strongly inhibited by the hydrophobic bilayer core. For example, in order for an ion to passively cross a lipid bilayer it has to leave a medium with a high dielectric constant of about $\epsilon = 80$ and venture into a hydrocarbon medium with a low dielectric constant of around $\epsilon \sim 1$–3. This amounts to an enormous electrostatic barrier of the order of $100\,k_{\rm B}T$. This barrier is very important for the cell to maintain the proper electrochemical potential across the membrane.

Nevertheless, ions can pass through the lipid bilayer, and the lipid bilayer structure and organization are determining factors for the degree of permeability. This is where the lipid phase transitions and phase equilibria come in. In Fig. 12.1a is shown the data for the passive permeability of a small negative ion, $S_2O_2^{2-}$, through a lipid bilayer of DMPC. The remarkable observation is that the lipid phase equilibria have a strong effect on the leakiness of the bilayer. At the phase transition of the pure lipid bilayers and at the temperatures corresponding to the phase lines in the phase diagram of binary mixtures (cf. Fig. 9.7), the permeability is anomalously large. Moreover, in the phase separation region of a lipid mixture, the mixed bilayer is quite leaky. These observations are fairly generic and have also been found for other ions like sodium, as shown below. The leakiness at the transitions is directly related to the small-scale structure and the lipid domains that develop as a consequence of the lipid phase transitions (cf. Chap. 11). The small-scale structure implies that the bilayer has a significant amount of defect and lines of defects through which the permeants can leak.

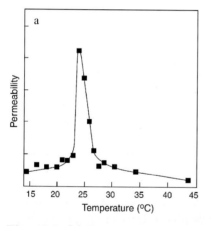 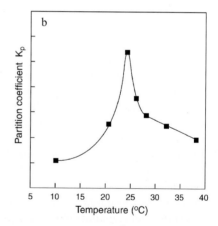

Fig. 12.1. (a) Passive permeability of a small negative ion, $S_2O_2^{2-}$, through lipid bilayers of DMPC. (b) Binding of ethanol to DMPC lipid bilayers. The binding is given in terms of the partition coefficient, K_p, which is a measure of the concentration of ethanol in the bilayer in relation to that in water

The second example demonstrates that foreign compounds that interact with membranes can sense the lipid phase transition. Figure 12.1b shows the binding of a simple alcohol, ethanol, to DMPC lipid bilayers. As is well known and more elaborately described in Sect. 17.2, ethanol has a strong effect on biological membranes, in particular, those of nerve cells. The figure shows

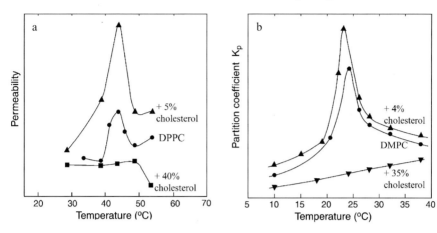

Fig. 12.2. (a) The effect of cholesterol on the passive permeability of sodium ions through DPPC lipid bilayers. (b) The effect of cholesterol on the binding of ethanol to DMPC lipid bilayers

that the partitioning of ethanol into lipid bilayers is strongly enhanced in the transition region. This is where the membrane gets really thirsty.

These two examples illustrate that a lipid bilayer becomes vulnerable in its transition region. It gets leaky and can be invaded by foreign compounds. There are many more known examples of dramatic events that become facilitated in the lipid phase transition region: membrane proteins are more easily inserted, cholesterol can more readily be exchanged between membranes, and the probability of membrane fusion and fusion of vesicles with lipid monolayers becomes enhanced.

Obviously, it is not desirable for biological membranes to be as vulnerable to nonspecific invasion of foreign compounds, as illustrated in Fig. 12.1. Eukaryotes have found a way of dealing with this problem by incorporating cholesterol into their plasma membranes. As shown in Fig. 12.2, large amounts of cholesterol serve both to suppress the anomalous permeability behavior as well as act to inhibit the binding of ethanol. It is a peculiar observation that small amounts of cholesterol have the opposite effect, leading to a softening of the bilayer and, hence, enhanced permeability and ethanol binding. This effect is caused by the special properties of the lipid-cholesterol phase diagram at low cholesterol concentrations, cf. Fig. 9.10.

12.2 Repelling Membranes

It is expected that the occurrence of the lipid phase transition makes the softness of lipid bilayers strongly dependent on temperature in the transition region. This is clearly borne out of the data for the bending and area compressibility modules shown in Fig. 12.3. Both quantities are seen to become anomalously low in the transition region. Hence, lipid bilayers become softer at their respective phase transition. The softening is more pronounced the shorter the fatty-acid chains are. This effect is an example of so-called thermal renormalization of the mechanical properties of the membrane. It is a consequence of the fact that, even though the main lipid phase transition is a first-order transition, it is subject to strong fluctuations. The fluctuations are stronger the shorter the lipid chains are. Formally, this can be expressed by saying that the lipid main phase transition is driven toward a critical point as the chain length is diminished, in much the same way as the boiling of water can be tuned toward a critical point by increasing the pressure. The fluctuations in the bilayer phase transition are manifested as local variations in the density and the thickness of the bilayer. One can think of this as wrinkles on the bilayer that tend to make it softer. These fluctuations are closely related to the lateral membrane heterogeneity described in Sects. 11.1 and 11.2.

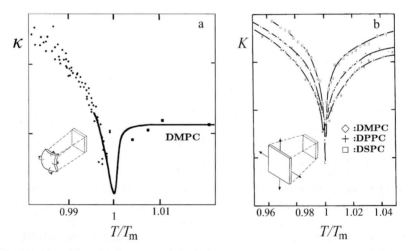

Fig. 12.3. Lipid bilayer membranes are softened near their phase transition. (**a**) The bending modulus, κ in (5.3), as a function of temperature for DMPC liposomes measured in the neighborhood of the phase transition temperature, T_m. (**b**) Area compressibility modulus, K in (5.2), for DMPC, DPPC, and DSPC bilayers calculated around their respective phase transition temperature, T_m. Both κ and K are given on a logarithmic axis

In Sect. 5.3, we remarked that the interaction between soft interfaces is influenced by an entropic force due to the undulations of the interfaces. According to (5.7) this entropic force depends on the bending modulus, κ, of the interface. Since lipid bilayers can become very soft in their transition region, one would expect that for a stack of lipid bilayers, as shown in Fig. 12.4a, the

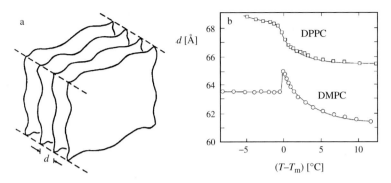

Fig. 12.4. (a) A stack of lipid bilayers subject to repulsive undulation forces leading to an inter-bilayer separation d. (b) The inter-bilayer distance as a function of temperature for two different types of lipid bilayers, DMPC and DPPC, in their transition region

softer the bilayers get the more they tend to repel each other and the further they will be apart, assuming that all other forces remain unchanged. This is, in fact, what one finds, as shown in Fig. 12.4b. Whereas the inter-layer distance, d, varies smoothly with temperature in the two lipid phases outside the transition region, there is a dramatic variation in the transition region. The overall drop in d going from the solid-ordered phase to the liquid-disordered phase is due to a thinning of the bilayer, as illustrated in Fig. 9.4. However, the thinner DMPC bilayer displays a peak in d at the transition. This reflects the stronger softening of the short-chain lipid that is closer to a critical point than DPPC.

The softening of lipid bilayers due to phase transition phenomena is the underlying universal mechanism behind the anomalous behavior of a large number of different membrane properties and phenomena: the peak in the specific heat, the minima in the mechanical modules, the peak in the passive permeability, and the enhanced binding of solutes and drugs like alcohols. All these phenomena are sensitive to lipid bilayer fluctuations. In Sect. 12.3 below we shall see that this universality also carries over to the action of certain enzymes.

12.3 Enzymes Can Sense Membrane Transitions

It was discussed in Sects. 5.4 and 11.2 that the main phase transition of lipid bilayers can be used as a mechanism to induce small-scale lateral heterogeneity in terms of lipid domains in bilayer membranes. This type of heterogeneity is rich in defects and can in fact be used to modulate the activity of certain enzymes that function at the surface of membranes. A prominent example is phospholipase A_2, which breaks down lipids as described in detail later in Sect. 18.1.

Figure 12.5 shows that the enzymatic activity of secretory phospholipase A_2 is strongly enhanced in the transition region of bilayers of different phospholipids. Moreover, the activity is stronger the shorter the fatty-acid chains are. Hence, the enzyme activity correlates to the lateral density fluctuations of the lipid bilayer. These fluctuations and the corresponding degree of lateral heterogeneity and lipid-domain formation are more pronounced the shorter the fatty-acid chains are, as illustrated in the top panel of Fig. 12.5. Alternatively, one could say that the softer the lipid bilayer becomes, as described in Sect. 5.4, the more vulnerable it becomes to attack by phospholipases. The same systematics is found when lipid bilayers are made softer by using lipids with increasing degree of unsaturation.

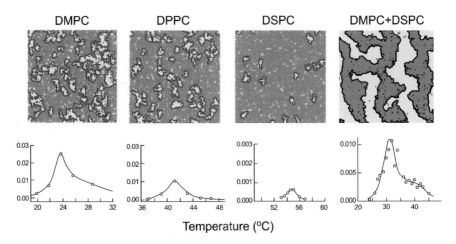

Fig. 12.5. The *bottom* panel shows the activity of secretory phospholipase A_2 as a function of temperature for three different lipid bilayers, DMPC, DPPC, and DSPC, in the neighborhood of their respective main phase transition. The activity is measured as the inverse lag time of the reaction, i.e., in units of sec^{-1}. For each lipid there is a strong increase of enzyme activity as the transition temperature is approached. The shorter the fatty-acid chains of the lipids the stronger the activity of the enzyme. Also shown is the activity of phospholipase A_2 as a function of temperature for a 1:1 binary mixture of DMPC-DSPC. The activity has maxima at the phase boundaries. The top panel shows typical configurations obtained from the computer simulation calculation of the lateral structure of lipid bilayers in the four cases. In the case of DMPC, DPPC, and DSPC, the lateral structure corresponds to temperatures just above the respective main transition. In the case of the DMPC-DSPC mixture, a possible lateral structure of the mixture in the phase-separation region is shown

In light of these findings it is not surprising that the activity of phospholipase A_2 is also strongly enhanced at the temperatures corresponding to the phase boundaries of binary lipid mixtures. This is demonstrated in Fig. 12.5

in the case of a 1:1 DMPC-DSPC mixture to which the phase diagram shown in Fig. 9.7 applies. Interestingly, the enzyme activity varies similarly to the passive transmembrane permeability in Fig. 12.1a, suggesting that the activity of the enzyme and the permeability are sensitive to the same type of underlying lipid-bilayer heterogeneity.

Hence, it appears that secretory phospholipase A_2 is an enzyme that is extremely sensitive to the physics of its substrate, i.e., lipid bilayers, their cooperativity and small-scale structure. The extreme sensitivity of phospholipase A_2 to the physical properties and small-scale structure of lipid bilayers can be exploited to facilitate targeted liposome-based drug delivery at diseased sites that are characterized by increased concentration and activity of phospholipases. This is in fact the case at cancerous and inflamed tissue, which we shall return to in Sect. 20.4.

12.4 Lipid Thermometer in Lizards

Calotes versicolor is a lizard whose body, like other amphibians, assumes the same temperature as its environment. Since quite large temperature variations can occur in the habitat of the lizard, it is obvious that the lizard needs to regulate the lipid composition of those membranes that are critical to its life functions. The same is true of animals that hibernate. It is essential that the membranes stay in their liquid phase in order to function.

An experiment has been conducted that suggests that *Calotes versicolor* use a lipid phase transition to monitor the ambient temperature. In this experiment, three populations of lizards were kept at three different temperatures: $16°C$, $26°C$, and $36°C$. After having acclimatized, the lizards were decapitated and the lipids of a certain part of their brains, the hypothalamus, were extracted and mixed with water to form lipid bilayers. The phase transitions of the three different extracts corresponding to the three different populations were then examined by measuring the specific heat as a function of temperature, as for pure lipids (cf. Fig. 9.3). The results are shown in Fig. 12.6.

Figure 12.6 shows that the specific heat in each case has a pronounced peak, signalling a lipid phase transition. Furthermore, the position of the peak occurs in each case at a temperature that is slightly below the temperature where the corresponding lizard population has been acclimatized. Hence, in response to temperature changes, the lizard regulates the lipid composition of its hypothalamus in order to remain in the liquid phase but very close to the lipid phase transition. Since lipid extracts from other parts of the brain do not exhibit the same kind of systematics, these results suggest that the lizard exploits the lipid phase transition as a kind of thermometer. This

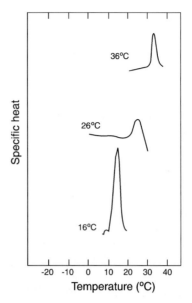

Fig. 12.6. Specific heat of lipid bilayers formed by lipid extracts from the hypothalamus of the lizard *Calotes versicolor*. Results are shown for three different cases corresponding to three populations of lizards that have been living at the temperatures given. The specific heat displays a peak at a temperature, which in each case is slightly below the temperature where the corresponding lizard population has been acclimatized

suggestion is in accordance with the fact that the hypothalamus is known to be the thermoregulatory part of the brain.

This example is an excellent illustration of how living systems adapt to environmental stress factors and how lipid diversity serves to provide for homeostatic control.

13 Proteins at Lipid Mattresses

13.1 Coming to Terms with Lipids

Proteins are the workhorses of the cell. They are involved at almost every stage of biological activity. Some of the proteins are enzymes that facilitate biochemical processes. Others are little motor molecules that make our muscles work. Still others take care of communication and transport of energy and matter. Proteins come in many different sizes and types. Some are fully integrated in membranes, while others are water soluble and float in the cell fluids or are attached peripherally to membranes.

However important proteins may be, most of them must, at some stage in their work, come to terms with the fact that they operate in an environment whose structure and dynamics are marshaled by lipids. Lipids grease the molecular machines of the cell, and they provide the compartments of the cell and its separation from the outside. They present themselves to the proteins as barriers as well as carriers and targets for protein function.

The mapping of genomes for whole species has highlighted the key role of those proteins that function in close association with membranes. More than 30% of the genome codes for integral membrane proteins. The fraction of proteins that bind peripherally to membranes is expected to be at least as large. Consequently, most of the genome is likely to code for proteins that function in relation to membranes. The interaction of these proteins with lipids is therefore crucial for their function. In addition, many drugs are proteins or peptides that act at membranes or at receptor molecules that are attached to membranes. The mode of action and the potency of these drugs depend on how they come to terms with the lipids.

Due to the low solubility of lipids in water, the most common situation for a peptide or protein working in a cell is not to meet a lipid on its own, but rather to encounter crowds of lipids in the form of a membrane or a small vesicular lipid capsule. Hence, it is not only the chemical character of the lipid that determines its interaction with the protein, but also the physical and social properties of the crowd to which the lipid belongs. So what is it that the proteins have to come to terms with?

In Part I of this book, we learned that lipids in the form of bilayers have a number of peculiar properties. Lipid bilayers are soft and can easily bend. Although very thin, the lipid bilayers have a finite thickness and a distinct trans-bilayer profile with built-in curvature stress. Moreover, due to the cooperative behavior caused by many lipid molecules being assembled together, the bilayer can sustain strong fluctuations and phase transition phenomena, leading to a heterogeneous lateral structure in terms of domains and rafts. We have already seen examples of how these conspicuous properties have distinct consequences for the barrier properties of the bilayer and the way the bilayer interacts with and binds other molecules.

In the present chapter, we shall describe in general terms how the physical properties of lipid bilayers on the one side govern the way peptides and proteins bind to and insert into membranes, and how the proteins and peptides on the other side modify the properties of the lipids. We shall focus on proteins and peptides that are either tightly bound to the lipid bilayer surface or are fully integrated and spanning the bilayer, as illustrated schematically in Fig. 13.1.

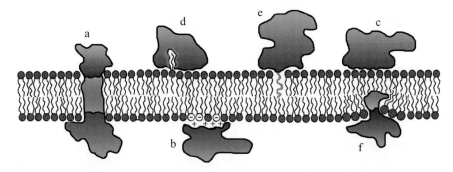

Fig. 13.1. Schematic illustration of an integral membrane protein that spans the bilayer (**a**) compared to five modes of binding a peripheral protein to a membrane surface: (**b**) electrostatic binding; (**c**) nonspecific binding by weak physical forces; (**d**) anchoring via a lipid extended conformation; (**e**) anchoring by a fatty-acid chain anchor attached to the protein; (**f**) amphiphilic protein partially penetrating the bilayer

There are three fundamental and interrelated constraints that an integral membrane protein has to conform to when imbedded in a lipid bilayer. Firstly, there is the finite thickness of the bilayer and the fact that the bilayer is amphiphilic. The hydrophobic membrane-spanning domain of the protein has to adapt to this by hydrophobic matching to the lipids, as will be described in Sect. 15.1. Secondly, there is the lateral pressure profile, which implies that the lateral pressure exerted by the lipids on the trans-membrane part of the protein varies dramatically through the bilayer, as shown in Fig. 8.2. Thirdly, there is the possibility of a built-in curvature stress in the membrane caused by certain lipids that have a propensity for forming non-lamellar lipid phases such as the inverted hexagonal phase H_{II} illustrated in Fig. 4.4. The curvature stress exerts a strain on the protein. The strain may be locally released by a conformational change in the protein. In the sections of the present chapter, we shall demonstrate how lipids can be in charge of protein organization and function by these three constraints.

The trans-bilayer profile in Fig. 8.1 and the lateral pressure profile in Fig. 8.2 describe the unusual interfacial properties that proteins and peptides are challenged by when approaching, penetrating, inserting, or translocating across a membrane. The regions denoted (1) and (2) in Fig. 8.1, which refer to the perturbed water layer and the chemically very heterogeneous mixture of hydrophilic and hydrophobic groups, are of particular importance in this context. These two regions, which together account for about half of the bilayer thickness, represent a chemical milieu that is very different from both the aqueous phase as well as the hydrophobic core of the membrane. These regions together are large enough to accommodate a whole α-helix lying parallel to the membrane surface. Peptides and proteins that insert into these regions are likely to change their structure and, hence, possibly their function.

But the lipids are also affected by the encounter with proteins. The main reason for this is that the lipid bilayer is like a soft mattress or cushion. The presence of proteins, therefore, leads to a number of effects, including changes in membrane thickness, changes in the conformational order of the lipid chains, and possibly molecular reorganization at the lipid-protein interface. It can be anticipated that the effects exerted by a single protein on the bilayer may propagate through the bilayer over distances where they influence other proteins, hence facilitating lipid-mediated indirect protein-protein interactions.

13.2 Anchoring at Membranes

Several molecular mechanisms can be imagined when water soluble peptides and proteins associate with the surface of membranes, and a host of forces of different origin can be involved. Some of these forces are weak and nonspecific physical forces, others are strong and long-ranged electrostatic forces, while still others involve formation of hydrogen bonds or even strong chemical bonds. In addition to these forces comes the possibility of a hydrophobic force that may drive an amphiphilic protein with a hydrophobic domain onto the membrane surface via the hydrophobic effect, as illustrated in Fig. 13.1f. We shall return to this possibility in Sect. 17.3, where we consider anti-microbial, amphiphilic peptides that bind to membrane surfaces.

Let us now illustrate how proteins and peptides anchor peripherally at membrane surfaces by a couple of the examples pictured in Fig. 13.1b–e. We already mentioned in Sect. 11.4 that a positively charged protein like cytochrome c, which is a peripheral membrane protein that transports electrons in the membranes of mitochondria, requires negatively charged lipids in order to bind to the membrane (Fig. 13.1b), and that the binding characteristics reflect the formation of domains of charged lipids underneath the protein. Hence, the cooperativity in the lipid-domain formation phenomena influences the binding of cytochrome c. Many other charged proteins are known to behave similarly.

An example of a protein that can bind to neutral membrane surfaces, such as DMPC bilayers, with weak physical forces (Fig. 13.1c) is the enzyme phospholipase A_2, which we discussed in Sect. 12.3 and shall return to in Sect. 18. Since this enzyme also carries positive charge, its binding can be enhanced on membranes with negatively charged lipids. The strength of the binding can be measured directly on the level of a single enzyme molecule by a special force spectrometer that is based on the same principle as atomic force microscopy.

The principle is illustrated in Fig. 13.2. The protein of interest is attached chemically to the tip of the cantilever of the force spectrometer by a linker molecule. The tip is rammed into the surface and withdrawn until the protein

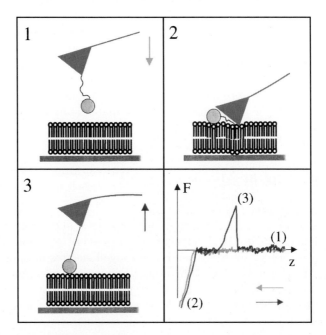

Fig. 13.2. Schematic illustration of the principle in single-molecule force spectroscopy applied to measure the binding of a protein to the surface of a membrane. The protein of interest is attached chemically to the tip of the cantilever of the force spectrometer by a linker molecule. The tip is rammed into the surface (1,2) and withdrawn (3) until the protein detaches from the membrane. The force, F, measured as a function of the distance, z, of the tip from the membrane surface is shown in the bottom right-hand frame. The maximal force is a measure of the strength of the binding

detaches from the membrane. The force is then measured as a function of the distance of the tip from the membrane surface upon retraction of the tip from the point of contact. The maximal force is a measure of the strength of the binding. Experiments of this type have demonstrated that the binding of phospholipase A_2 is weaker to electrically neutral lipid bilayers of DMPC than to negatively charged lipid bilayers made of DMPG. Obviously, the difficulty in successfully performing such a molecular fishing experiment relies on the delicate chemistry of attaching the protein to the tip. Furthermore, the quantitative interpretation of the force measurements is complicated by the fact that the measured force depends on how fast the protein is pulled off the surface.

Our last example involves a small synthetic and artificial polypeptide with only ten amino acids. The poly-peptide is acylated by a saturated hydrocarbon chain with fourteen carbon atoms. This chain can be used by the poly-peptide to anchor to the membrane surface, as illustrated in Fig. 13.3, which is an example of the prototype in Fig. 13.1e. The polypeptide is folded like a hairpin because it contains in the middle a particular amino acid (proline) that induces a turn. Furthermore, it has another particular amino acid (tryptophan) near the fatty-acid chain anchor. Tryptophans are known to be abundant in natural membrane proteins at the protein domains that locate in the hydrophilic-hydrophobic interface of the membrane.

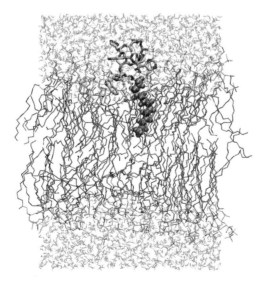

Fig. 13.3. Computer simulation of a DPPC lipid bilayer incorporated with an acylated poly-peptide (C_{14}-His-Trp-Ala-His-Pro-Gly-His-His-Ala-amide). The water soluble polypeptide exhibits a hairpin conformation outside the membrane, and the hydrophobic fatty-acid chain anchor is buried in the membrane core

The strategy of using a hydrocarbon chain as a membrane-anchoring device is used by a large number of natural membrane proteins, e.g., those proteins called lamins which provide the scaffolding at the inner side of the membrane that bounds the cellular nucleus in eukaryotes.

13.3 Spanning the Membrane

Proteins of the membrane-spanning type shown in Fig. 13.1a constitute a huge class of some of the most important functional proteins in biology. Examples include channels, pore complexes, pumps, and receptors. Despite their diversity in actual structure and function, they have one architectural motif in common, which is a membrane-spanning domain with a length that is limited to narrow range. This range is dictated by a physical constraint imposed by the thickness of the lipid bilayer membrane in which the protein is embedded. In this context, it is striking to note that the membrane-spanning domain of membrane proteins is the evolutionarily most conserved amino-acid sequence, hence suggesting a universal principle operative for lipid-protein interactions.

The long poly-peptide chain that constitutes an integral membrane protein can be thought of as a thread sewn through the lipid bilayer once or several times. The transmembrane domain, therefore, consists of one or several stretches of amino acids, as illustrated in Fig. 13.4. Water soluble proteins are folded in a way that arranges for the hydrophobic amino acids to be inside, away from the water and the hydrophilic amino acids on the outside facing the water. In contrast, membrane-spanning proteins are inside-out such that the transmembrane part has hydrophobic amino acids facing the lipid membrane and sometimes hydrophilic or charged amino acids toward their

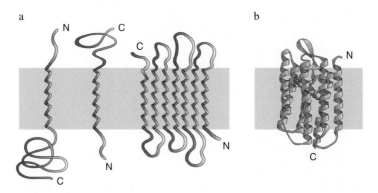

Fig. 13.4. (a) Schematic illustration of the universal structural motif of integral membrane proteins with one or several turns that pass through the lipid bilayer. (b) Three-dimensional structure of bacteriorhodopsin, which is a trans-membrane protein with seven α-helices spanning the lipid bilayer

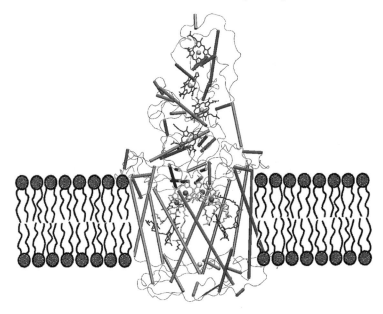

Fig. 13.5. The photosynthetic reaction center protein from the membranes of the bacterium *Rhodopseudomonas viridis*. This protein, which uses light to transport protons across the bacterial membrane, was the first membrane protein whose structure was resolved in atomistic detail. The protein and the lipids in the bilayer are not drawn to scale

interior. In addition, transmembrane proteins often have very large parts outside the membrane, on either side, as illustrated in Fig. 13.5. Being in water, these extremities, which quite often carry the functional units of the proteins, are arranged like water soluble proteins. Dictated by lipids, integral membrane proteins are therefore bound to be amphiphilic, i.e., hydrophobic in the middle and hydrophilic in the two ends.

An example of the folded structure of an actual integral membrane protein in a lipid bilayer environment is given in Fig. 13.4b, which shows the protein bacteriorhodopsin that functions as a light-driven proton pump in certain salt-loving bacteria. This protein has seven transmembrane α-helical polypeptides spanning the membrane. In fact, this protein is, together with a small number of other integral membrane proteins, the only membrane proteins for which a detailed three-dimensional structure with atomic resolution is presently available.

Bacteriorhodopsin represents a special case in which the protein spontaneously forms two-dimensional crystals in its natural bacterial membrane. Examples of such crystals are shown in Fig. 13.6. Using single-molecule techniques similar to those described above in Sect. 13.2, a team of German researchers led by Hermann Gaub and Daniel Müller succeeded in extracting a single bacteriorhodopsin molecule from a two-dimensional crystalline

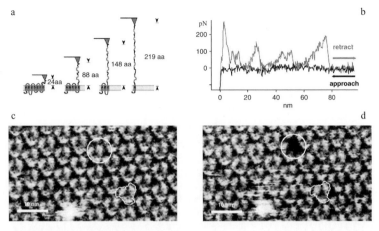

Fig. 13.6. Extracting a single integral membrane protein from of a membrane. A single bacteriorhodopsin molecule (**a**) sitting in a two-dimensional crystal (**c**) in a membrane is pulled on by the tip of an atomic force microscope. The molecule is attached to the tip in the –COOH end of the protein. The resulting force versus distance curve is shown in (**b**). Frame (**d**) exposes a hole in the crystal inside the circle where a single bacteriorhodopsin molecule has been pulled out

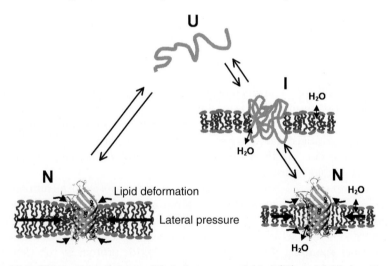

Fig. 13.7. Illustration of the equilibria between unfolded (U) and folded native (N) states of outer membrane protein A (OmpA) of Gram-negative bacteria. OmpA forms an eight-stranded β-barrel in the N-terminal transmembrane domain. Denatured OmpA in the aqueous phase spontaneously refolds into lipid membranes upon removal of the denaturants. Correct spanning of the hydrophobic domains across the membrane is a major determinant of the folding-unfolding equilibria and the folding pathways. In the case of too thin membranes, an intermediate state (I) is found in which the protein is inserted but not correctly folded

array. The process of extraction is a kind of reverse folding process. The measured force curve in Fig. 13.6b can be interpreted in terms of successive unfolding of the seven transmembrane helices of the protein. The same team of researchers have recently succeeded in observing the reverse process of refolding the protein into the membrane.

As part of their natural cycle, integral membrane proteins have to undergo the reverse process of the one illustrated in Fig. 13.6a. When the proteins are transported and secreted through the cell, e.g., on their route from the site of synthesis to where they are determined to do their work, they have to be inserted into, secreted through, and correctly folded into the membrane. During these processes, the hydrophobic parts of the protein have to come to terms with and span the membrane. This process is illustrated in Fig. 13.7.

Part III

Lipids in Action

14 Cholesterol on the Scene

14.1 Molecule of the Century

Cholesterol was discovered in 1815 by the French chemist Michel E. Chevreul, who found it in human gall stones. Its precise molecular structure, shown in Fig. 2.5a, remained unknown, however, until 1932. In the following decades, the biosynthetic pathway to cholesterol was worked out, and during the 1970s and 1980s, the relationship between the molecular evolution of sterols and the evolution of species was unravelled. The actual regulation of the cholesterol biosynthesis in humans by low density lipoprotein receptors became clarified in the last quarter of the twentieth century. All this important work led to three Nobel Prizes: the 1927 Chemistry Prize to Heinrich O. Wieland for his work on cholesterol structure, the 1964 Physiology and Medicine Prize to Konrad Bloch for his work on cholesterol synthesis, and the 1985 Physiology and Medicine Prize to Michael S. Brown and Joseph L. Goldstein for their work on regulation of cholesterol biosynthesis. In total, fourteen Nobel Prizes have been awarded to sterols or topics related to sterols. Research on cholesterol was undoubtedly a key issue in the last century.

If cholesterol has had high priority as a research area, it has drawn an even higher attention among the public. The reason for this is that cholesterol is known to be related to the number one killer of Western populations: coronary heart disease and atherosclerosis. It is also associated with adiposis, which presents an increasing problem worldwide. Therefore, cholesterol is probably the lipid that has the worst reputation. This is somewhat of a paradox, since cholesterol is both an important structure builder in all cells of our body in addition to being an important metabolite and source of important vitamins and hormones. Few realize that cholesterol is the single most abundant type of molecule in our plasma membranes, accounting for 30–50% of the lipid molecules, and that all eukaryotic organisms on Earth use similar amounts of cholesterol (or related sterols) in their plasma membranes.

So cholesterol is an absolutely essential lipid for the higher forms of life. We shall in this chapter advocate the viewpoint that part of cholesterol's success in life is due to its unique capacity for imparting to lipid membranes some very special physical properties. And we shall learn that nature has taken great care and has spent a long time evolving this unique molecule.

14.2 Evolutionary Perfection of a Small Molecule

Lipids are possibly some of the oldest organic molecules on Earth. As mentioned in Sect. 1.1, lipids or other interface-forming molecules are required for forming the capsules that can protect enzymes and genes from a hostile environment. Obviously, life, as it evolved since its first appearance on Earth about 3.8 billion years ago, used the molecules available, either those already existing or the new ones that were produced by various living organisms. The life-forms that were fit would survive according to the Darwinian selection principles. Cholesterol or related higher sterols like ergosterol and sitosterol, cf. Fig. 2.5, were not available for a very long time for the very reason that the chemical conditions for the biochemical synthesis of these higher sterols were not there. What was lacking was molecular oxygen.

For convenience, we shall in the following only refer to cholesterol that is pertinent to animal life. In a discussion of plants and fungi, cholesterol should be replaced by sitosterol and ergosterol, respectively. The differences in molecular structure of these three higher sterols, as shown in Fig. 2.5, are of minor importance for the general arguments of the present chapter.

In order to appreciate how the advent of cholesterol released new driving forces for the evolution of higher organisms, it is instructive to study the variation of molecular oxygen (O_2) in the atmosphere of Earth since its creation almost five billion years ago. Figure 14.1 correlates the oxygen partial pressure in the atmosphere with major events in the evolution of life. Before the evolution of the blue-green cyanobacteria that can produce O_2 by photosynthesis, the partial pressure of O_2 was exceedingly low, possibly as low as one into ten million of an atmosphere. It increased gradually to concentrations that were large enough to support life-forms that exploit oxygen by respiration, possibly around 2.8 to 2.4 billion years ago. Up until then, eubacteria and archaebacteria (prokaryotes) were the only forms of life. But along with the availability of molecular oxygen, eukaryotic life appeared on the scene. From then on, as the oxygen pressure was rising, there was a proliferation of eukaryotic diversity. This suggests that there is a conspicuous coincidence between the emergence and rise of the eukaryotes and the availability of molecular oxygen.

It has been proposed that the availability of molecular oxygen removed a bottleneck in the evolution of species and that the crucial molecular entity in this process is cholesterol (and related sterols). This proposal is supported by the fact that eukaryotes universally contain high concentrations of cholesterol in their plasma membranes, whereas cholesterol is universally absent in prokaryotes. It should be added that the internal membranes of eukaryotes carry very low concentrations of cholesterol. There is a striking gradient in the cholesterol concentration from the mitochondria (3 weight%), over endoplasmic reticulum (6 weight%), to the Golgi (8 weight%). As we shall return to in Sect. 15.1, this gradient follows the secretory pathway of proteins through the cell towards the plasma membrane, indicating that cholesterol may be

14.2 Evolutionary Perfection of a Small Molecule

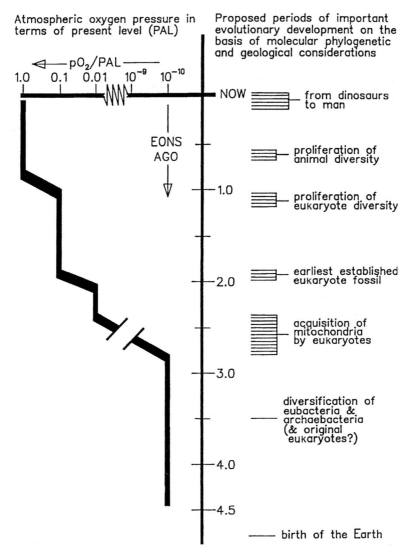

Fig. 14.1. Variation over time of the partial pressure of molecular oxygen, p_{O_2}, in the atmosphere of Earth relative to present-day level. The time is given in billions of years. Note the logarithmic scale for the pressure. To the *right* is shown important evolutionary developments of life based on information from molecular phylogenetic and geological considerations

involved in the sorting of proteins. The fact that mitochondria have almost no cholesterol is in line with Lynn Margulis's symbiosis theory according to which the mitochondria in eukaryotes are ancient prokaryotes that were engulfed by the eukaryotes to take care of the respiratory process.

In order to understand the background for the provocative statement about cholesterol and its unique role in evolution, we have to draw upon the fundamental work by Konrad Block who worked out the biochemical pathway for synthesis of cholesterol. Figure 14.2 illustrates this pathway, ranging from squalene to lanosterol to cholesterol. The path starts with the linear molecule squalene, which becomes cyclizised into the characteristic steroid ring structure. Konrad Block showed that there is no plausible way of cyclizing squalene in the absence of oxygen, and it is even more unlikely, if not impossible, to perform the next steps that lead from lanosterol to cholesterol. These steps can be seen as a successive streamlining of the hydrophobic surface of the sterol by removing from one to three of the methyl ($-CH_3$) groups that protrude from the flat face of the molecule. Chemical evolution in the absence of molecular oxygen along the sterol pathway would therefore have to stop with squalene. Konrad Block has termed the oxidative process leading to cholesterol "the evolutionary perfection of a small molecule" and thereby pointed out that not only genes changed during evolution, but so did lipids and, in particular, sterols.

The significant difference in molecular smoothness between lanosterol and cholesterol can be seen in Fig. 14.3. The three additional methyl groups on lanosterol make this molecule rougher and bulkier than cholesterol. It is surmised that Darwinian evolution has selected cholesterol for its ability, via its smoothness, to optimize certain physical properties of the membranes. It is, however, unclear which physical properties are the relevant ones in this context. Moreover, it is uncertain which amount of optimization cholesterol can provide.

A clue to these questions may come from considering the contemporary biosynthetic pathway to cholesterol as the living "fossil" of the evolutionary pathway to cholesterol. Along this pathway, lanosterol is a precursor to cholesterol. In other words, the temporal sequence of the biosynthetic pathway could be taken to represent the evolutionary sequence. The evidence for this viewpoint, which appears highly convincing, comes from Konrad Bloch's studies of sterol biochemistry and organism evolution. The concept of living molecular fossils offers a framework for a research program geared toward identifying the physical properties that are relevant to evolutionary optimization, without having to face the impossible problem of performing experiments on evolutionary time scales.

14.3 Cholesterol Fit for Life

One of the most conspicuous properties of cholesterol in relation to the physical properties of lipid bilayers is its ability to stabilize a very special membrane phase, the liquid-ordered phase. As described in Sect. 9.4, the liquid-ordered phase is a proper liquid in the sense that it allows for the necessary rapid diffusion in the plane of the membrane and, at the same time, supports

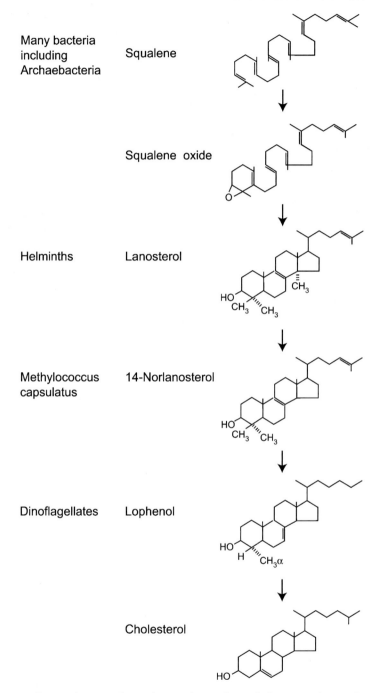

Fig. 14.2. Biosynthetic pathway for synthesis of sterols from squalene to lanosterol to cholesterol. To the *left* are indicated organisms that use the molecular precursors to cholesterol

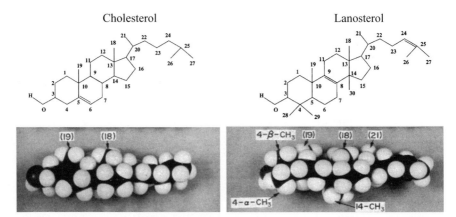

Fig. 14.3. Chemical structures and space-filling models of cholesterol and lanosterol highlighting lanosterol's extra three methyl groups that lead to a hydrophobically less smooth surface of the molecule

a high degree of conformational order in the lipid chains. Cholesterol acts as a kind of anti-freeze agent. This leads to mechanically stable membranes that are thicker and less leaky than the same membranes without cholesterol. The question then arises whether the stabilization of the liquid-ordered phase is something peculiar to cholesterol. And if so, could it be that the specific physical property of membranes, which possibly has been optimized during evolution, is in fact the liquid-ordered phase?

Whereas for obvious reasons we cannot answer the second question, the first question can be subject to an experimental test using model systems. Figure 14.4 shows the differential effects on lipid-chain ordering going from lanosterol to cholesterol at a temperature corresponding to liquid membrane phases. Cholesterol is clearly better than lanosterol to order the lipid molecules. Larger ordering leads to thicker and less leaky membranes. In this sense, cholesterol presents an advantage over lanosterol.

The physical reason why lanosterol induces less ordering is related to its less smooth steroid skeleton. As a result, lanosterol is incapable of producing a liquid-ordered phase in lipid bilayers. This is clearly demonstrated by the phase diagrams in Fig. 14.5, which compares the phase equilibria for cholesterol with that of lanosterol. The molecular evolution from lanosterol to cholesterol can therefore be pictured as an evolution in the structure of the phase diagram from one with no liquid-ordered phase to one where a liquid-ordered phase is stable over a substantial range of temperatures and sterol concentrations. One of the advantages of the presence of the liquid-ordered phase is that it may be required for forming membrane rafts, as discussed in Sects. 11.3 and 11.4. A particularly interesting observation is that rafts, and hence the functions controlled by rafts, are suppressed when cholesterol is replaced by lanosterol. Lanosterol is a poor raft-former.

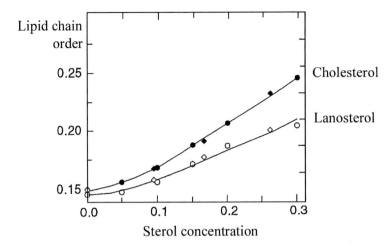

Fig. 14.4. Ordering of lipid chains in a lipid bilayer induced by lanosterol and cholesterol as a function of sterol concentration

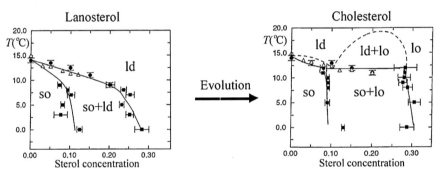

Fig. 14.5. Evolution from lanosterol to cholesterol seen as an evolution in the phase equilibria toward a situation with a stable liquid-ordered membrane phase. The labels on the different phases correspond to the liquid-disordered (**ld**) phase, the solid-ordered (**so**) phase, and the liquid-ordered (**lo**) phase. The sterol concentration is given in mole%

A large number of studies have shown that many biological functions in eukaryotic cells become impaired if cholesterol is replaced with lanosterol. As an example, insects are unable to synthesize sterols and have to get cholesterol via their diet. If only provided with lanosterol, they die. Another example is yeast functioning under anaerobic conditions. Yeast will thrive on cholesterol, but cannot survive on a diet of lanosterol.

It is possible that prokaryotes, which generally lack sterols, have developed other strategies to impart to their membranes the strength and mechanical stability that cholesterol assures in eukaryotic plasma membranes. An example is provided in the work by the French chemist Guy Ourisson, who points to a class of molecules called triterpenes and hopanoids, which are bacterial lipids. These molecules are rather stiff and may, hence, provide for mechanically strong membranes. Despite the fact that these molecules constitute the greater amount of the biomaterial on Earth, very little is known about their effects on the physical properties of membranes.

14.4 Cholesterol As a Killer

Having said all the good about cholesterol and its key role in the evolution of higher life, a few words are in order about nutrition and the transport and turnover of cholesterol in our bodies. Irregularities in this transport system is at the heart of cardiovascular diseases and atherosclerosis. Our source of cholesterol is twofold: Some gets synthesized in the endoplasmic reticulum and some is supplied by the diet. It is known that the biosynthesis of cholesterol is a regulated process in the sense that uptake of dietary cholesterol inhibits biosynthesis.

Cholesterol is transported in the bloodstream by little particles called *low-density lipoproteins* (LDL). These particles, which are no more than 20 nm in diameter and contain less than two thousand molecules, are packages of cholesteryl esters (cholesterol linked to a fatty acid) wrapped in a layer of phospholipids together with a protein called apolipoprotein B-100. In addition to cholesterol, LDL particles transport other water insoluble compounds, such as vitamins and hormones, to various cell types.

The LDL particles are secreted from the liver in the form of larger precursor particles (very-low-density lipoproteins). The LDL particles are removed from the bloodstream by a particular membrane-bound receptor called the LDL receptor, which was discovered by Joseph L. Goldstein and Michael S. Brown for which they were awarded the Nobel Prize in 1985. Upon binding to the receptor, the LDL particle is internalized in the liver cells by endocytosis and then taken up and degraded by lysosomes that turn the cholesteryl esters into free cholesterol. When the level of free cholesterol then rises, transcription is suppressed of the genes coding for LDL receptors as well as those coding for the enzymes that are involved in the synthesis of cholesterol. Fewer LDL receptors lead to a negative feedback loop that increases the level of LDL in the blood. The regulation of the biosynthesis of cholesterol is thereby closely linked to the clearing of cholesterol from the blood.

How does this make cholesterol a potential killer? If for some reason the level of LDL in the blood is too high, the LDL particles may deposit their load of fatty acids from the cholesteryl esters and the phospholipids at the walls of the arteries. The trouble is then that some of these fatty acids, in particular, arachidonic acid, together with the fatty acids already present in the artery wall can produce an inflammation of the walls. The deposition of fatty acids lead to plaques that eventually can provoke blood clotting, angina, and heart attacks. Consequently, cholesterol can in a very indirect and complex manner cause serious diseases.

There is a particular monogenetic disorder called familial hypercholesterolemia whose incidence is at least 0.2% of the population in Western countries. This disorder causes LDL to accumulate in the blood and leads to atherosclerosis. Familial hypercholesterolemia is caused by a genetic defect that leads to a deficit of LDL receptors. There are three other known monogenetic disorders that disrupt the fine regulation between cholesterol synthesis and blood clearance by LDL.

If LDL particles are the bad guys, there is also a set of good guys called *high-density lipoproteins* (HDL). HDL also carries cholesterol and takes it from the various body tissues and delivers it to the liver. In the liver, cholesterol is degraded into bile salts and subsequently excreted from the body. Obviously, the balance between LDL and HDL levels in the blood is important for a healthy condition.

In recent years, a number of drugs have emerged that help to lower cholesterol levels in the blood. The more promising of these drugs is a class of compounds called statins. The statins operate by blocking an enzyme in the synthetic pathway to cholesterol. There is some evidence that statins reduce heart attacks and prolong life.

It is an issue of much controversy as to what extent dietary fats, in particular, cholesterol and the ratio of unsaturated fatty acids to saturated fatty acids, influence the incidence of coronary diseases and atherosclerosis. This is not the place to enter this controversy. A few remarks are in order, however, to indicate the complexity of the problem. Recent studies have indicated that there may be little relationship between dietary cholesterol and coronary diseases. Some argue that although the incidence of heart diseases has not dropped during years of public ban of a fatty diet, the mortality rate has declined mainly because better medical treatment has become available, which, in turn, enhances the chances of surviving a heart attack. There is some indication that an intake of polyunsaturated fatty acids, in particular, linoleic and α-linolenic acids, as discussed in Sect. 16.1, provides for longevity, possibly by lowering the level of LDL in the blood.

15 Lipids in Charge

15.1 Lipids and Proteins Match Up

The physical constraint imposed on integral membrane proteins by the lipid bilayer thickness, as illustrated in Fig. 13.4, suggests that a mechanical hydrophobic matching principle may be operative. Hydrophobic matching means that the hydrophobic length of the transmembrane domain is matched to the hydrophobic thickness of the lipid bilayer. There are obvious energetic advantages of matching. Therefore, hydrophobic mismatch could be a controlling mechanism for the way proteins interact with lipids in membranes.

Some important implications of the hydrophobic matching principle can be gauged from the sketch in Fig. 15.1a, which shows a situation where the hydrophobic thickness of a lipid bilayer is smaller than the hydrophobic length of the transmembrane protein. In order to compensate for the mismatch, the soft lipid bilayer yields and the lipid molecules closest to the protein stretch out to cover the hydrophobic core of the protein. This leads to a perturbed region around the protein. For a lipid bilayer with a single lipid species, this perturbed region is characterized by a larger average lipid bilayer thickness and a higher conformational chain order. Since lipid bilayers and membranes under physiological conditions are liquids, the perturbed region is a statistical entity in the sense that lipids diffuse in and out of the region. The physics of the situation is similar to that of water wetting the inside of a glass. An illustration of bacteriorhodopsin in a membrane to which it is hydrophobically well matched was shown in Fig. 8.6.

Figure 15.2 shows an example of a case in which an integral membrane protein is positioned in a lipid bilayer that is slightly too thick. The protein is an aquaporin, which is a water transporting protein that also facilitates the transmembrane transport of glycerol in the plasma membrane of *Escherichia coli*. The protein is basically a pore that allows some molecules to pass through the membrane and not others. The pore is here placed in a POPC lipid bilayer, and the consequences of hydrophobic matching is seen as a local thinning of the membrane around the protein. The lipid chains near the protein contract to fulfill the matching condition. Water channels of this type are also responsible for maintaining the osmotic balance over the cell membranes in the human body. In total, these molecular pores transport

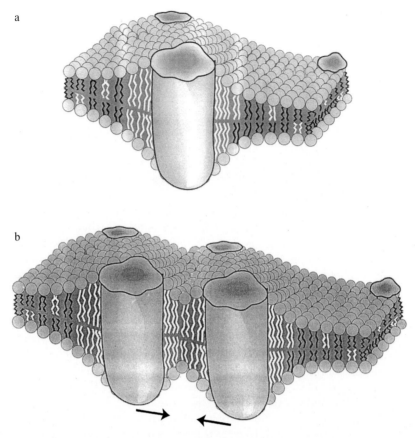

Fig. 15.1. Hydrophobic matching principle for lipid-protein interactions in membranes. (a) An integral protein in a too thin membrane. (b) Lipid-mediated attraction between two mismatched integral proteins

every day almost two hundred liters of water across the membranes in our bodies.

Direct evidence for the hydrophobic matching of the lipids to the hydrophobic size of the aquaporin channel is provided in Fig. 15.3, which shows the lipid bilayer thickness profile in the neighborhood of the protein. A thinning of the bilayer is observed around the protein, corresponding to a specific lipid annulus containing 40–80 lipid molecules that are substantially influenced by the protein. Hence, the lipid-protein interactions furnish a differentiated membrane region which is related to the formation of membrane rafts discussed in Sect. 11.3.

For lipid bilayer membranes with several different lipid species, the hydrophobic matching principle furnishes an even richer behavior. The perturbed region around the mismatched protein could imply a local de-mixing

Fig. 15.2. Hydrophobic matching leads to a thinning of a POPC lipid bilayer incorporated with an aquaporin transmembrane protein whose hydrophobic domain is too short to match the hydrophobic bilayer thickness. The picture is obtained from a Molecular Dynamics calculation on a model with full atomistic details

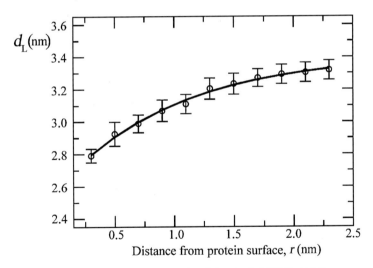

Fig. 15.3. Hydrophobic thickness profile, $d_L(r)$, of a POPE lipid bilayer around an aquaporin transmembrane protein, cf. Fig. 15.2. Hydrophobic matching leads to a thinning of the bilayer around the protein. The data is obtained from a Molecular Dynamics calculation on a model with full atomistic details

of the lipid molecules such that the lipid species that provides for the better hydrophobic match is recruited at the lipid-protein interface. By this mechanism, proteins can perform a sorting of the lipids, leading to an annulus around the protein. Since the lipid molecules are subject to diffusion, this annulus is a statistical entity. Its lipids would generally exchange with the lipids outside the annulus, although specific electrostatic or chemical binding may occur. It is clear that the selection of which lipids are accumulated near the protein can be varied by changing membrane composition, thermodynamic conditions, or by adding compounds, like drugs, that modify the energy stored in the mismatch. More importantly, the annulus can change if the protein undergoes conformational changes that lead to a change in the matching condition. In this way, the hydrophobic matching principle provides a direct link between lipid-bilayer properties on the one hand and protein structure and function on the other.

If we then consider the situation with more than one protein shown in Fig. 15.1b, the possibility arises of two or more proteins sharing the perturbed region of lipids. This would be energetically favorable and, therefore, lead to an effective attraction between proteins. This attraction is mediated by the lipids and their cooperative behavior and is like a colloidal force. In this way, the membrane acts as an elastic sheet or mattress whose elastic deformation energy, upon intercalation of the proteins, can be minimized by reorganizing the lateral distribution of the proteins in the plane of the membrane. The full consequences of this scenario are illustrated in Fig. 15.4, where changes of the thermodynamic conditions in a model lipid bilayer with two different lipid species of different chain length are seen to drive a random protein dispersion to a state of aggregation and crystallization. The organization principle is pure hydrophobic matching. An application of this principle to a specific system is illustrated in Fig. 15.5, which shows the formation of two different kinds of two-dimensional crystals of bacteriorhodopsin in the plane of the membrane.

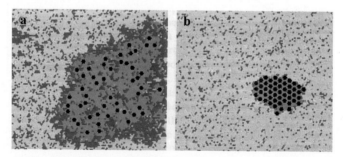

Fig. 15.4. Snapshots from computer simulation calculations on a model for a binary lipid mixture of lipids with two different chain lengths. By varying the temperature, protein crystals can be formed (**b**) or dissolved (**a**). The proteins are shown as solid hexagons

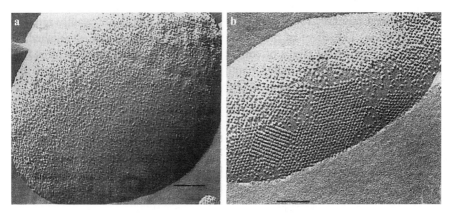

Fig. 15.5. (a) Dispersion and (b) crystallized domains of the transmembrane protein bacteriorhodopsin in lipid membranes. The scale bar is 100 nm

The range over which the integral proteins influence the lipid bilayer depends on a number of details such as degree of mismatch, lipid composition, and temperature. This range is related to the coherence length and average lipid domain size introduced in Sect. 11.1, and it can vary from a single layer of lipids around the protein to very many. In some sense one can say that the proteins pick up or harvest the lipid fluctuations and domains in the lipid bilayer. This, in turn, leads to a stabilization of the domains. The range over which different proteins can "feel" each other through the lipid bilayer is also set by the coherence length. Hence, the lateral organization of proteins in a lipid bilayer can to some extent be modulated by altering the coherence length, e.g., by changing temperature or by adding specific substances such as drugs that will change the coherence length.

The finer details of the hydrophobic matching principle have been studied intensively in model membrane systems with synthetic amphiphilic polypeptides. These peptides can be specifically designed to span the membrane, and they can be synthesized with different lengths of the hydrophobic domain in order to vary the mismatch.

It appears to be a general finding that proteins in their natural membrane are well matched to the membrane's hydrophobic thickness, i.e., the thickness of the physiologically relevant liquid membrane phase. Almost all integral membrane proteins stop functioning when the membrane is taken into the solid phase. This makes sense, considering that the bilayer thickness is substantially larger in the solid phase. In the solid phase, lateral diffusion is slowed down at least by a factor of hundred. This is another important reason why membranes stop functioning when taken into solid phases.

By these statements we have already suggested that the hydrophobic matching principle may provide a mechanism for coupling lipid membrane properties to the functional state of proteins. There is a vast amount of experimental evidence which strongly suggests that the hydrophobic matching principle is relevant for membrane organization as well as for a variety of membrane functions. For example, it has been found for a number of membrane channels, ion pumps, and sugar transporters that they, when incorporated into lipid bilayers of different thicknesses, function optimally for a certain narrow range of thicknesses, where they presumably are hydrophobically well matched. Thickness alterations induced internally or by external stimuli may therefore be seen as a way of triggering these proteins to enhance or suppress their function, as illustrated schematically in Fig. 15.6.

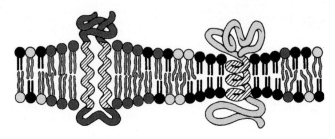

Fig. 15.6. Schematic illustration of triggering the function of an integral membrane protein by changing the hydrophobic mismatch

As an example, we show in Fig. 15.7 how the activity of two different integral membrane proteins that pump ions across membranes depends on the thickness of the membrane. Ca^{2+}-ATPase is important for muscle cell action, whereas Na^+-K^+-ATPase takes care of the delicate balance of sodium and potassium ions across membranes. Na^+-K^+-ATPase is responsible for using about one third of all the energy our bodies turn over. The figure shows that the activity of both ion pumps is maximal for a certain lipid type and, hence, for a specific membrane thickness. If cholesterol is added to the membrane, the data in Fig. 15.7b demonstrate that the maximum moves toward lipid membranes made of shorter lipids. This can be rationalized via the hydrophobic matching principle, recalling that cholesterol tends to thicken fluid membranes, thereby compensating for the shorter lipids.

This latter observation suggests a more general principle to be operative by which cholesterol may be used as a regulator of membrane function and the sorting and targeting of proteins, possibly via hydrophobic matching. The following serves as an illustration. Proteins are synthesized at the ribosomes placed in the endoplasmic reticulum. From there they are transported via the Golgi to the various parts of the cell where they belong, e.g., at the plasma membrane. This transport, which is referred to as the secretory pathway, requires a sorting of the proteins, which is partly performed in the Golgi. Some proteins carry specific tags that will actively target them to their destination, others will passively flow through the cell.

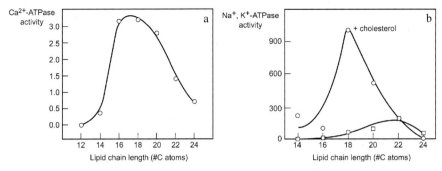

Fig. 15.7. (a) Activity of the membrane-bound enzyme Ca^{2+}-ATPase as a function of the hydrophobic thickness of the lipid bilayers in which it is incorporated. The hydrophobic thickness is given by the number of carbon atoms of mono-unsaturated PC lipids. The activity exhibits a clear maximum. (b) Activity of the membrane-bound enzyme Na^+, K^+-ATPase as a function of the hydrophobic thickness of the lipid bilayers in which it is incorporated. When cholesterol is incorporated (here in the amount of 40%), the maximum is moved toward membranes made of shorter lipids

The question arises as to how these flowing proteins end up in the right membranes. It has been proposed that the sorting along the secretory pathway may be performed by means of a gradient in the hydrophobic thickness of the membrane systems that the proteins have to pass on their way to their target. Indeed, the amounts of cholesterol and sphingomyelin, which both tend to enlarge membrane thickness, are found to increase going from the endoplasmic reticulum, via the Golgi, to the plasma membrane. Furthermore, there is evidence that the proteins that are supposed to stay in the Golgi have hydrophobic domains that are shorter by about five amino acids compared to those of the plasma membrane. The set of different membranes along the secretory pathway may hence act as a molecular sieve, exploiting the hydrophobic matching condition. It is possible that this sieving mechanism is controlled by the membrane rafts discussed in Sects. 11.3 and 11.4. Since cholesterol has a significant effect on membrane thickness and since integral membrane proteins are hydrophobically matched to their membranes, it is likely that the transmembrane proteins have closely coevolved with the sterols. The proposed sorting mechanism has recently been challenged by data that shows that the bilayer thickness of exocytic pathway membranes is modulated by membrane proteins rather than cholesterol.

15.2 Stressing Proteins to Function

Closer inspection of Figs. 13.1 and 15.1 suggests that although membrane surface properties and bilayer hydrophobic matching may be important for the way peripheral and integral membrane proteins interact with membranes, we may have left out something important. As we learned in Chaps. 4 and 8, lipid molecules have effective shapes, they may display propensity for forming non-lamellar phases, and they are subject to an awful lot of stress due to the lateral pressure variation across the bilayer. The way proteins perturb lipids on the one side and the way lipids exert stresses on the proteins on the other side would have to involve these features. This becomes even more clear when asking how lipids can influence the functioning of integral membrane proteins, or how lipid structure may influence the binding of peripheral proteins.

Although being two sides of the same problem, we shall for convenience first discuss consequences of non-lamellar lipids on lipid-protein interactions, and then describe how the lateral pressure profile relates to protein function.

A substantial part of the lipids found in natural membranes are very poor bilayer-formers and in fact have a propensity for forming non-lamellar structures like the inverted hexagonal phase, H_{II} (cf. Sects. 4.2 and 4.3). This, of course, does not mean that these biological membranes are non-lamellar. Rather, it means that the lipids have to be together with specific proteins in the membrane in order to make the lamellar state favorable. However, it also means that these membranes carry an intrinsic curvature stress field and an inherent instability toward curved structures. This instability must have some advantage for function. One could imagine the instability to be locally released in connection with protein binding, protein insertion, membrane fusion, and conformational changes in the cycle of protein functions. Some of these possibilities are illustrated in Fig. 15.8.

Let us start with a peripheral membrane protein in order to see how this could work: protein kinase C, which is one of the most important enzymes involved in the signal transduction system of the cell. Upon stimulation of the cell by, e.g., neurotransmitters, hormones, and growth factors, protein kinase C becomes activated upon binding to the plasma membrane, leading to a complicated cascade of biochemical signals that eventually influence cell growth, cell differentiation, as well as exocytosis. Requirements for binding to the membrane and, hence, for activation of the enzyme are lipids with acidic head groups, like PS^-, and the presence of calcium ions. Calcium ions require water for solvation, and, in the competition for water, the membrane surface becomes dehydrated, leading to a larger curvature stress.

It has been proposed by the Finnish biophysicist Paavo Kinnunen that this curvature stress could be released if some lipid molecules assume an extended chain conformation by flipping one of the fatty-acid chains to the outside of the membrane. This flip would normally be energetically very costly because of the hydrophobic effect. However, if the chain can be accommodated in a putative hydrophobic crevice in a protein, as illustrated in Fig. 15.8a, it will not only release the curvature stress of the membrane but, at the same time, facilitate the membrane anchoring and, hence, the activation of the enzyme. Protein kinase C has such a hydrophobic crevice. The presence

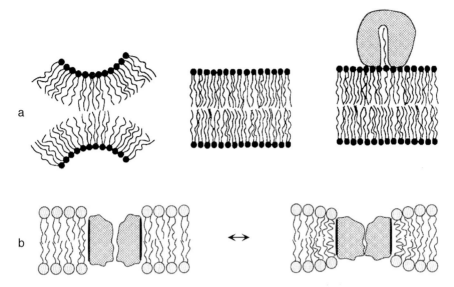

Fig. 15.8. Schematic illustrations of lipid-bilayer structures subject to curvature stress and the possible influence of curvature stress on protein and membrane function. (**a**) Hypothetical situation of two relaxed lipid monolayers with intrinsic curvature corresponding to an inverted hexagonal phase (H_{II}) (*left*), the corresponding bilayer with built-in curvature stress (*middle*), and the case of partially released curvature stress by binding to a peripheral protein via the extended chain-conformation mechanism proposed by Paavo Kinnunen (*right*). (**b**) Effect of curvature stress on the opening of a membrane channel. In this case, amphiphilic molecules with a big head group and a small tail tend to close the channel, since they have a propensity for forming H_I phases. In contrast, amphiphilic molecules with a small head and a large tail tend to open the channel, since they have a propensity for forming H_{II} phases

of the extended chain anchorage is further supported by the finding that the addition of PE lipids enhances the enzyme activity. PE lipids have a small head group and hence display a propensity for forming H_{II} phases (cf. Sects. 4.2 and 4.3). This further increases the curvature stress and promotes the formation of the extended lipid chain conformation that, in turn, explains the enhanced activation of protein kinase C. There are also some indications that the mechanism of anchoring via a lipid-extended conformation may play a role in the binding of cytochrome c as well as a number of other membrane active proteins that indeed have hydrophobic pockets to accommodate the extended lipid chain.

We then turn to a more complex situation that involves the integral membrane protein rhodopsin, which is a seven-helix transmembrane protein similar to bacteriorhodopsin in Fig. 13.4. Rhodopsin is the light-sensitive protein in the visual pigment of our retina that upon activation of light initiates the signalling pathway that eventually leads to vision. An essential stage of this process involves a certain transition between two conformational states of rhodopsin, the so-called M-I and M-II states. The M-II state is believed to correspond to a more elongated form of rhodopsin than the M-I state. The transition, therefore, implies a change in hydrophobic mismatch. Studies have shown that the M-II state requires the presence of lipids that have a propensity for forming H_{II} phases. The so-called retinal rod outer segment membranes, in which rhodopsin functions, are known to have almost fifty percent of the polyunsaturated fatty acid docosahexaenoic acid (DHA) (Fig. 2.2c), which, due to the many double bonds, indeed supports curved structures. The fact that the M-I to M-II transition can be activated by other non-lamellar forming lipids such as PE lipids suggests that it is the physical curvature stress release by the lipids, rather than a specific chemical reaction between DHA and rhodopsin, that is the controlling mechanism. Additional support for this viewpoint is the finding that short alcohol molecules, which are known to position themselves in the hydrophobic-hydrophilic interface of the membrane and hence counteract the stability of the H_{II} phase, can de-activate the transition. The question then remains as to why the visual system as well as our brains, as discussed in Chap. 16, have chosen to use the rather special polyunsaturated lipid DHA for manipulating the curvature stress in the neural membranes. Despite its obvious importance, this question remains unsolved at present.

The American physical chemist Robert Cantor has put the relationship between curvature stress and protein function on a more quantitative footing by presenting a simple mechanistic model picture based on how the lateral pressure profile, cf. Fig. 8.2, can couple to protein function via the stresses it exerts on a transmembrane protein. This picture, which is illustrated in Fig. 15.9, relates the work, W, required to induce a transition between two states, r and t, of the protein, characterized by two different cross-sectional area profiles, $A_r(z)$ and $A_t(z)$, of the protein, to the lateral pressure profile, $\pi(z)$,

$$W = -\int_z \pi(z)[A_t(z) - A_r(z)]\mathrm{d}z \ . \tag{15.1}$$

The crucial point here is that the protein needs to have a non-cylindrical shape in order to sense the lateral pressure profile and the possible changes in the profile under the influence of other factors. Estimates of the amount of work required to change the conformational state of an integral membrane protein suggest that the stress changes that the lipid bilayer can provide indeed should suffice to activate the protein. In fact, it has recently been demonstrated by detailed calculations already referred to in Sect. 15.1 and in Figs. 15.2 and 15.3 that the capacity of aquaporins to transport water becomes affected by the hydrophobic matching condition and variations in the lateral pressure profile induced by changing the size of the lipid polar head group.

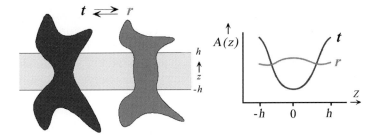

Fig. 15.9. Schematic illustration of the change in cross-sectional area profile, $A(z)$, of an integral membrane protein that undergoes a conformational transition and a shape change in a lipid bilayer

The fact that many integral membrane proteins seem to require non-lamellar lipids for their function, taken together with the observation that most natural membranes contain large amounts of non-lamellar lipids, may provide a clue to overcoming a serious obstacle in membrane biology. The trouble is that the full three-dimensional structure has only been worked out in atomic detail for very few integral membrane proteins compared to the large number of known structures for water soluble proteins. This reflects a somewhat paradoxical situation, since, as alluded to above, membrane-associated proteins may be the largest class of proteins judging from the maps of the human genome. However, the thousands of protein structures that have been worked out to date are almost all water soluble. The reason for this is that integral membrane proteins, due to their amphiphilicity, are very difficult to crystallize, and crystals are prerequisites for solving protein structure by X-ray methods. Membrane proteins denaturate when they are taken out of their membrane environment. Therefore, all sorts of tricks have to be played in order to form crystals of membrane proteins. These tricks often involve the protection of the hydrophobic transmembrane domain of the protein by various detergents and lipids. Interestingly, it has been found that small crystallites of some integral membrane proteins can be produced in membrane-mimicking systems made of non-lamellar-forming lipids that stabilize the cubic lipid structures shown in Fig. 4.4. The mechanism for producing the crystals is proposed to be related to the bicontinuous nature of the cubic phase that would allow the proteins to diffuse freely to the nucleation sites of the crystal formation.

15.3 Lipids Opening Channels

A particularly elegant and quantitative way of studying the effect of curvature stress on the opening and closing of membrane channels has been pioneered by Olaf Sparre Andersen, who has used a small polypeptide, gramicidin A, as a model protein. Gramicidin A is an antibiotic that forms dimers in lipid membranes, typically by joining two monomers back to back, as described later in Sect. 17.4 (cf. Fig. 17.3a). The dimer conducts small positive ions, and the activity of single channels in membranes can be measured by electrophysiological techniques. The gramicidin channel can be seen as a simple model for the more complex opening and closing of a membrane channel molecule or carrier, shown in Fig. 15.8b. Obviously, the propensity for forming dimers and, hence, for activating the model protein, depends on hydrophobic matching and, in the case of a mismatch, on how well the lipids can adopt to a locally curved interface toward the dimer.

Studies of this model system have not only shown that a good hydrophobic match enhances dimer formation, but have also clearly demonstrated that in the case of a mismatch, where the bilayer is too thick to accommodate the dimer, the formation of dimers can be facilitated by adding lipids that have a propensity for forming curved structures. These lipids presumably help to mediate the curvature stress that otherwise would build up at the lipid-peptide interface. In thick bilayers, gramicidin A has been found to induce fully developed H_{II} phases. An interesting corollary to these observations is that gramicidin A can be used as a molecular force transducer that can be exploited to measure the elastic stresses not only in model membranes, but also in biological membranes where gramicidin A is introduced in very small amounts and where its channel activity is measured by electrophysiological techniques.

There is a particular class of membrane proteins, the mechano-sensitive channels, that have evolved to facilitate ion conductance in response to a stress exerted by the membrane. These proteins are nano-machines that work as transducers of mechanical strain dissipated from the membrane. The most well studied example is the bacterial large conductance mechano-sensitive channel (MscL) from *Escherichia coli* shown in Fig. 15.10. Hydrophobic matching and curvature stress concepts can be used to interpret the experimental data for the conductance and how it varies when non-lamellar lipid species are incorporated into the membrane. MscL is a helix bundle protein, and experiments as well as Molecular Dynamics simulations have supported a mechanism for channel opening that involves an iris-like expansion of the conducting pore.

15.4 Lipids Mediate Fusion

Fusion of membranes are important processes in the functioning of all eukaryotic cells. In particular, the extended systems of transportation and trafficking of macromolecules within the cell and between cells involve the merging and separation of membranes, as illustrated schematically in Fig. 15.11. The molecular mechanism of membrane and vesicle fusion is still a somewhat controversial issue, and a number of models have been proposed for the transient arrangements of the lipid molecules during the process. The current picture

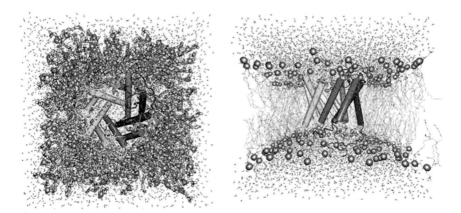

Fig. 15.10. Molecular Dynamics simulation snapshots of the bacterial large conductance mechano-sensitive channel (MscL) from *Escherichia coli*. *Top* and *side* views are shown corresponding to the beginning of the simulation where lateral tension is exerted on the channel

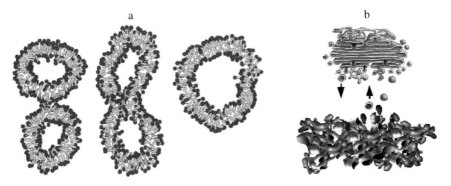

Fig. 15.11. (a) Molecular simulation of the fusion of two vesicles revealing the formation of a fusion intermediate in the form of a membrane stalk that only connects the outer monolayers of the two fusing vesicles. (b) Fusion and fission processes of transport vesicles that are trafficking proteins between the endoplasmic reticulum (*bottom*) and the Golgi apparatus (*top*)

involves the formation of a stalk intermediate, as illustrated in Fig. 15.11a. The membrane stalk is a neck-like structure in which only the outer monolayers of the two fusing membranes are connected. During the stalk formation, regions of high membrane curvature are formed, possibly mediated by local non-lamellar structures, cf. Fig. 4.4. At some stage, the two inner monolayers make contact and an aqueous pore is formed connecting the lumen of the two vesicles.

Transport across the plasma membrane is facilitated by endocytosis or exocytosis, by which material is secreted by the fusion of vesicles with the plasma membrane. Similarly, internalization of material from the outside into the cell can take place by invagination of the plasma membrane to form a vesicle that carries the material into the cell cytosol. The transport of a virus in a membrane envelope takes place by a similar mechanism. Nerve function relies on the fusion and fission of vesicles across the gap between neighboring nerve cells. In this process, the vesicles carry the neurotransmitter molecules which pass on the nerve signal.

The trafficking of proteins in the secretory pathway inside the cell is mediated by vesicles and relies on various fusion and fission processes, e.g., from the ER to the Golgi, as illustrated in Fig. 15.11b. While the ER constitutes a reticular membrane that spans almost the entire cell, the Golgi apparatus is a stack of distinct membrane cisternae that comprise different chemical milieus. After proteins have been synthesized in the ER, they accumulate in specialized membrane domains, the so-called ER exit sites, from where they are exported by means of small, coated membrane structures, called COPII vesicles. Upon reaching the Golgi apparatus, these vesicles release their protein cargo, which then is modified while being processed through the stack of cisternae. Finally, the products are either sent to their final destination, e.g., the plasma membrane of the cell or are recycled by means of other vesicles called COPI vesicles.

The transportation within the cell is often served by lysosomes and vesicles that are targeted by specific proteins. The lipid vesicles are small cargo-carrying packets that carry stuff from one compartment of the cell to another. The vesicles' ability to fuse with membranes is a prerequisite for many functions. The fusion processes involving membranes and vesicles are not fully understood on the molecular level, and several different mechanisms have been proposed. In particular, a number of fusion intermediates of the fusing lipid bilayers have been put forward. Two things are clear, however. Firstly, certain fusion peptides and proteins are involved, e.g., the so-called SNARE-proteins, proteolipid complexes, and receptors activated by calcium ions. Secondly, fusion requires a local rearrangement of the lipids in the involved membranes in order to allow for regions of very high curvature. In particular, the formation of inverted micellar intermediates, as indicated in Fig. 15.11a, may occur for topological reasons. Obviously, lipids with a large propensity for forming H_{II} structures, such as PE lipids and lysolipids, will facilitate fusion processes, whereas bilayer-forming PC lipids will not.

16 Being Smart – A Fishy Matter of Fat

16.1 The Essential Fatty Acids

You are what you eat. All the molecular building blocks that our body is made of are supplied from the diet. Our food consists of protein, sugar, and fat (in addition to a lot of important minerals). These food molecules are broken down, e.g., into amino acids and fatty acids, and put together again to produce precisely the kind of proteins and lipids that we need in order to build our cells and maintain their specific functions. In the case of lipids, our body has systems that are capable of transforming some fatty acids from the food into other fatty acids that are the ones needed for the construction of certain lipids. For example, it is common for animals to be able to transform saturated fatty acids into mono-unsaturated fatty acids with a double bond in position 9 along the carbon chain, whereas they lack the ability to make unsaturated bonds in positions 12 and 15. Only plants have the capability of doing so. The transformation in plants is facilitated by a host of enzymes, so-called elongation and desaturation enzymes, which can extend the length of fatty-acid chains and increase the number of double bonds (C=C) on the chains.

Since animals need to get these unsaturated fatty acids from their diet, they are referred to as *essential fatty acids*. The essential fatty acids for humans (and other vertebrates) are polyunsaturated and contain eighteen carbon atoms. The acids are called *linoleic acid*, C18:2n-6, and *α-linolenic acid*, C18:3n-3, and they have two and three double bonds, respectively. n refers to the position of the double bond nearest to the methyl end of the molecule. C18:2n-6 has two double bonds in positions 9 and 12, and C18:3n-3 has three double bonds in positions 9, 12, and 15. Their molecular structures are shown in Fig. 16.1.

From the two types of essential fatty acids, two families of polyunsaturated and super-unsaturated fatty acids can be formed by elongation and desaturation. The pathways for the biochemical processes of elongation and desaturation are depicted in Fig. 16.2. These two families are also called n-6 (or ω-6) and n-3 (or ω-3) fatty acids. Members of one family cannot always substitute in a given function for members of the other family. Furthermore, the human body does not have efficient systems to chemically transform compounds from one family into compounds from the other family. Mammals like

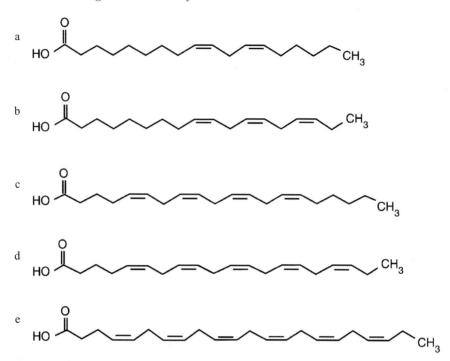

Fig. 16.1. The essential fatty acids: (**a**) linoleic acid, C18:2n-6, and (**b**) α-linolenic acid, C18:3n-3, and some super-unsaturated fatty acids derived from them according to the pathways in Fig. 16.2. (**c**) Arachidonic acid (AA), (**d**) eicosapentaenoic acid (EPA), and (**e**) docosahexaenoic acid (DHA)

man can produce super-unsaturated fatty acids like arachidonic acid (AA), docosapentaenoic acid (DPA), and docosahexaenoic acid (DHA) in the liver once they have got the polyunsaturated essential fatty acids. But the process is very slow and energy consuming. The question is then if the rate of production can keep up with the need. In Sect. 16.2, we shall discuss the hypothesis that in the evolution and development of the human neural system and the brain it has been absolutely critical that the diet contain sufficient amounts of super-unsaturated fatty acids in order for the development of the neural system and the brain to keep up with the growth of the body. From where can we then get these super-unsaturated fatty acids?

Linoleic acid, which is the precursor for the members in the n-6 family, e.g., AA, is found in large amounts in oils from various seeds such as sunflower, corn, and soybean. α-linolenic acid, which is the precursor for the members in the n-3 family, is synthesized only in higher green plants, algae, and phytoplankton. Since green plants and algae constitute the largest part of the biomass on Earth, α-linolenic acid is probably the most dominant fatty acid on Earth. The sources for AA and DHA are egg yolk, the meat and

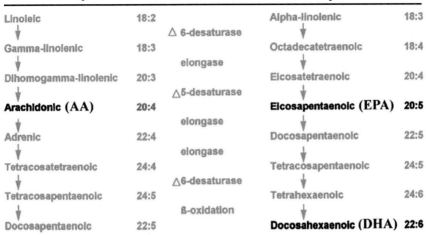

Fig. 16.2. Elongation and desaturation pathways for the essential fatty acids, linoleic acid, C18:2n-6, and α-linolenic acid, C18:3n-3. Some of the chemical structures are shown in Fig. 16.1. The three highlighted fatty acids, arachidonic acid (AA), eicosapentaenoic acid (DPA), and docosahexaenoic acid (DHA), are important for neural membranes and the brain. The docosapentaenoic acid (DPA) from the n-6 family is different from DPA from the n-3 family in that the first one is missing a double bond in positions 19 and 20 and the other one in positions 4 and 5

organs of animals, marine algae, as well as water-based animals, cold-water fish, and shell fish that directly or indirectly feed on algae. As an example, fatty fish are rich in DHA, e.g., almost 50% of the fatty acids in salmon is DHA, whereas it is only 0.2% in cow.

The essential fatty acids are also important as precursors for hormonal compounds. For example, linoleic acid is the basis for forming hormone-like molecules called *eicosanoids*, e.g., *prostaglandins* and *leukotrienes*, which are involved in human reproduction processes, blood flow, and the immune system.

16.2 Evolution of the Human Brain

The human brain is mostly fat; about 60% of the dry weight consists of lipids. Fifty percent of these brain lipids contain essential fatty acids and their derivatives. A very large part of these lipids is made of the long and superunsaturated fatty acids we discussed in Sect. 16.1. The fatty acid profile of the human brain is characterized by approximately equal proportions of AA and DHA from the two n-6 and n-3 families. For example, DHA represents up to about 20% of the fatty acids, and AA and DPA account for about 15%

and 5%, respectively. A similar dominance of super-unsaturated fatty acids is found in other neural tissue, e.g., in the visual system and the retina where DHA represents about 50% in the disk membranes of the rod outer segment. It is striking that whereas a diverse group of species like mammals, reptiles, and fish use rather different fatty acids in their muscles, livers, and other organs, the fatty acid composition of their brains is rather similar. Moreover, only long-chain super-unsaturated fatty acids are used for the brain, whereas mixtures of different chain lengths are used for other tissues.

During the last couple of million years, the line of bipedal primates that are believed to be the ancestors of man experienced a rapid growth in the cerebral cortex. The modern *Homo sapiens* is likely to have originated in Africa between 100,000 and 200,000 years ago. The British neuro-chemist Michael Crawford has presented an exciting hypothesis that the accessibility of DHA has been a determining factor in the evolution of the human brain. He takes his starting point in the observation that what distinguishes *Homo sapiens* from other mammals, even other primates, is its large brain. Or to be more precise, it is the combined facts that the human brain is large and at the same time the ratio between brain weight and body weight is large, too. To put this observation in perspective, it should be noted that the brain–to–body weight ratio for different species is generally found to decrease logarithmically with increasing body size. Small mammals like the squirrel, rat, and mouse have brain–to–body weight ratios around 2%, in chimpanzee it is 0.5%, in larger gorilla 0.25%, and it is below 0.1% in rhinoceros and cow. Some striking exceptions to this relation are humans (2.1%), dolphins (1.5%), and other animals that evolved and lived at the land-water interface. What is special for these animals is that they both have big brains and large bodies. In some way, the development of their brains has been able to keep up with the growth of their body size.

Michael Crawford now brings chemistry and nutrition in as an evolutionary driving force by hypothesizing that the evolution of the human brain could only have taken place where sources of DHA have been plentiful, i.e., at the seaside where marine food is available. There is some fossil evidence in support of this hypothesis. Animals that evolved on the savannah had to make do with the little DHA they could produce themselves or obtain by eating other animals and collecting their DHA. Carnivores should then have an advantage over herbivores. In fact, it is found that e.g., lions have higher levels of DHA than zebras and cows. However, the difference is not big enough to use as an explanation of why the brain of the omnivore and "killer-ape" *Homo sapiens* is bigger than the brain of the herbivore chimpanzee.

According to Crawford, it is the joint availability of AA (from the n-6 family) and DHA (from the n-3 family) which is the determinant of brain growth and function. In mammalian brains, the balance between AA and DHA is close to 1:1. Fish is a rich resource of DHA, e.g. the AA:DHA ratio is 1:40 in cod muscle membranes. Then what about marine mammals? In fact it is found, that the AA:DHA ratio in the brain, liver, and muscle tissue of dolphins is also close to 1:1. That is, dolphins are in a biochemical sense still land mammals that happen to live in a marine environment. Obviously, both AA and DHA are needed in large but comparable amounts in order to develop the large and complex neural systems like brains in humans and dolphins.

How do fish fit into this picture? Fish have ample access to DHA, and they carry plenty of DHA in their muscle fibers. Still, they have small brains compared to their body weight. Crawford's answer to this problem is that the fish embryo and the larvae during the critical phase of development of its neural system in fact do not have much DHA available. It has to do with what little was supplied in its egg. In contrast, mammals are supplied by AA and DHA from the placenta during their long gestation period. In fact, the access to these essential fatty acids during the early development of the fetus is so important that the human maternal brain may suffer a 3–5% reduction in the last trimester of pregnancy. Moreover, the newborn child continues to get plenty of AA and DHA from mother's milk, which contains high levels of AA and DHA. A lack of DHA in these early phases of development of the fetus' and child's neural system can imply that the child becomes mentally incapacitated and in worst cases suffers irreversible loss of vision and cognitive abilities. This suggests the importance of adding AA and DHA to infant formula for improving brain and visual development.

The question therefore remains as to what fish use their large storage of DHA and DPA for. Since they cannot use it for further development of their own neural systems, it must have some other still unknown biological function.

Several other questions await their answer. What is so special about DHA that makes it so uniquely important for the function of the brain and the visual system? Why is it that the more readily available DPA, which biochemically is not that different from DHA, cannot be used instead? The Canadian physicist Myer Bloom has proposed that DHA may optimize certain physical properties of membranes, specifically, mechanical properties that can promote optimum conditions for the functioning of certain membrane-bound proteins (G-proteins), membrane fragility that is needed for the plasticity of the brain, as well as electrical properties of importance for signalling in the brain. In Sect. 15.2, we remarked on the possible role of DHA as a non-lamellar-forming lipid and how its propensity for supporting H_{II} structures may have some bearings on the functioning of the light-sensitive protein rhodopsin in the retina, which can be considered the most forward part of the brain.

16.3 Lipids at the Border of Madness

A large brain is not enough to make *Homo sapiens* creative and intelligent. Neanderthals had larger brains than us, but are not thought to have been more intelligent. An interesting although very controversial hypothesis has been proposed by the late British scientist and medical doctor David Horrobin, who has suggested that the critical factor is connectivity of the brain, i.e., how capable the brain is of making micro-connections (synapses) between the dendritic extensions of the nerve cells (neurons). This is the point where phospholipids and the regulation of phospholipids by special enzymes come in.

A neuron can make thousands of to hundreds of thousands of connections with other neurons. Each synapse involves phospholipids, in particular, those containing AA and DHA. In developing the correct type of connectivity during the embryonic stage, it is critical that there is a tight control of the growth, the decay, and the regrowth of the developing synapses in their growth zones. This control requires intimate regulation of the metabolism of AA and DHA in the growth zones. For this purpose, a series of enzymes is needed, most notably acyl-transferases that put the fatty-acid chains on the glycerol backbone of the lipid, and phospholipase A_2 and phospholipase C that can remodel the fatty acids and the head groups of the lipid molecules, respectively. A number of other enzymes, co-enzymes, and lipoproteins are also likely to be involved. The lipoproteins are responsible for the delivery of fatty acids to the tissues. The evolutionary changes in the lipoproteins and plasma proteins that facilitated effective and rapid transport of fatty acids into the brain may have caused the depositing of fats in the human breast, the buttocks, and the subcutaneous adipose tissue. These human attributes distinguish *Homo sapiens* from the great apes.

If something fails in this tight regulation in the neural growth zones, anomalies may arise in brain development, possibly leading to psychiatric disorders. Horrobin mentions two major possibilities for breakdowns, each of them caused by at least a single gene. The first possibility is increased liberation of AA and DHA due to increased activity of one of the phospholipases that remodel the lipids. The other possibility is the reduced rate of incorporation of AA and DHA into the phospholipids due to decreased activity, e.g., in one of the acyl-transferases. The kind of disorders that may be the consequence of the first type of failure include manic depression. Serious results of the second type of breakdown in regulation include schizotypy, which is related to dyslexia.

The question then arises as to what will happen if there is more than one gene defect and both types of breakdown in the regulation occur simultaneously. Horrobin suggests that this is the biochemical source of schizophrenia. According to United Nations WHO–standardized criteria, 0.5–1.5% of a population, irrespective of race, will develop schizophrenia. It has been argued that schizophrenia is more heavily expressed in populations that adopt a Western-style diet, which is characterized by low levels of polyunsaturated fatty acids. Thus it is likely that the diet is just as an important factor as genetics in governing mental ill health.

Horrobin now makes the conclusion that the possibility of schizophrenia in humans comes about at the same time as the biochemical systems were developed to assure abundance of micro-connections in the brain, that is, at the same time humans became human. This is supposed to have happened between 50,000 and 200,000 years ago. Although having been large for a long time, only at that stage did the human brain become complex enough to be the playground for creativity, intelligence, and cultural imagination. In that sense, and according to Horrobin, schizophrenia is the illness that made us human.

17 Liquor and Drugs – As a Matter of Fat

17.1 Lipids Are Targets for Drugs

A large number of pharmacologically active drugs are hydrophobic or amphiphilic compounds, suggesting that their targets in the body are hydrophobic sites or at hydrophobic-hydrophilic interfaces. The hydrophobic sites could be either proteins and receptors or the interior of cellular membranes. The interfaces could be surfaces of membranes. In any case, the lipids are among the prime suspects, directly or indirectly. Even in the cases where lipids are not directly involved at the site of action, they are likely to be so at some stage during the route from administration and application of the drug until it finally arrives at the target. This holds true whether the drug is taken orally, injected into the blood, or applied through the skin. In all cases, there are tremendous lipid-dominated barriers for the drug molecules to overcome. Crossing the intestinal barrier, the blood-brain barrier, or the dermal barrier involves coming to terms with lipids in organized form, typically as lipid bilayers and cell membranes. The ability of drugs to pass the blood-brain barrier can sometimes be predicted based on the interface activity of the drug.

Some drugs may not make it that far, since they could be caught by the body's defense system and broken down by the various chemicals and enzymes in the body. Others may not even get started if they have too low solubility in the bodily fluids. In fact, many very potent and promising drug candidates never come into use for the very reason that they cannot be prepared in a formulation that makes them sufficiently water soluble. At this stage, lipids, rather than functioning as a barrier, could be used as formulating agents, in the form of an emulsion, or as drug carriers, in the form of a micelle or a liposome. We shall discuss liposomes as drug-delivery systems later in Chap. 20.

Finally, lipids come in as a target. In the case of drugs that have to be targeted to a receptor in a specific membrane, the surface properties of the lipid membrane in which the receptor is incorporated can play an active role. Furthermore, some drugs have lipids as their prime targets, e.g., alcohols and certain drugs used for anesthesia are believed to act at the lipids of neural membranes. Similarly, those potent peptides, which we use as antibiotics to kill bacteria by destroying their cell membranes, function by binding to the bacterial lipids. We shall discuss these two examples in more detail in Sects. 17.2 and 17.3.

The long list of drugs that strongly interact with lipid membranes includes general and local anesthetics, antipsychotics, antibiotics, anti-tumoral drugs, antidepressants, tranquilizers, antihistamines, antifungal compounds, and analgesics. The action of some of these can be enhanced by increasing their affinity to membranes. As an example, the potency of desmopressin, which is an antidiuretic hormone peptide that regulates the water drainage of the body, is known to be enhanced 250-fold by being kept in close contact with membranes. The trick used involves attaching two palmitic acid chains to the drug to anchor it in the membrane by the mechanism discussed in Sect. 13.2.

17.2 Alcohol and Anesthesia

Alcohol and anesthetics are widely used in our society – in bars, at the dentist, and in the hospital, and most people have come in contact with these drugs. Although most people know how they affect our nervous system and our behavior, it may come as a surprise that the molecular mechanism of action of alcohol (ethanol) and general anesthetics is not known. General anesthetics is a large and diverse class of chemical compounds, of which ethanol is one. The class also includes halogenated alkanes and volatile substances like ethers, heavy rare gases, and nitric oxide ("laughing gas"). General anesthetics, in contrast to local anesthetics, function on the central nervous system, and there appears to be no relationship between chemical structure and potency. Another striking observation is that the clinical concentration of any active compound needed to induce anesthesia is about 2%, basically independent of which organism one talks about, from tadpole to man.

It has been known for a century, since the days of Meyer and Overton, that the potency of general anesthetics correlates well with their solubility in olive oil (or lipid) relative to that in water. This relationship is given by the partition coefficient

$$K = x_{\text{lipid}}/x_{\text{water}} , \qquad (17.1)$$

where x refer to the anesthetics concentration in the two media. This remarkably simple relationship is still the basis for clinical use. As a curiosity, the relation is also the basis for rationalizing why women with the same body weight as men tolerate less alcohol, the reason being that men contain on average 13% more water than women. Another remarkable observation is the so-called pressure-reversal phenomenon, which refers to the finding that general anesthesia can be reversed by applying hydrostatic pressure. Obviously, the partitioning in (17.1) is shifted toward the aqueous phase at elevated pressures. Hydrophobic pressure is the only known antagonist to anesthesia, and the reversal process is fully reversible.

The Meyer-Overton relation and the pressure reversal of anesthesia seem to put lipids and fats at the center of the problem. It suggests that the potency of anesthetics is related to their hydrophobicity or amphiphilicity, and that the site of action in some way must be related to a hydrophobic site at the neural membranes. The question is *which* site.

There are basically two schools. One who subscribes to the viewpoint that there are specific receptors or protein-binding sites for the anesthetics, and another one that claims that the anesthetic effect is mediated by the lipids in the neural membranes. It appears, however, that it is generally agreed that the site of action of general anesthetics is related to ion channels in the neural membranes. This, in turn, raises the question whether the action on this site is direct, at the channel or indirect via the lipid membrane.

There is ample evidence that ethanol and many general anesthetics have a strong effect on the physico-chemical properties of lipid bilayers. They lower the main phase transition temperature as well as in some cases make the membrane more unstable toward forming cubic lipid structures. One of the troubles with these effects is that they are hardly detectable at those anesthetic concentrations that are applied in the clinic. This appears to rule against the lipid school. However, this may not be so obvious as it seems, since the local concentration of the drug in domains, rafts, or at the interface between the lipid and the membrane proteins may be considerably different from the global concentration of the drug in the membrane. These local effects can be caused by the cooperative nature of the lipid bilayer, which is highly affected by the presence of small interfacially active molecules. In some cases, these molecules tend to enhance the fluctuations and stabilize lipid domains by accumulating at the domain boundaries. This leads to a heterogeneous distribution of the drugs, with regions of substantially higher local concentration than the average global membrane concentration. As a consequence, protein assemblies and aggregates in lipid domains can become affected by the drug.

A clue to understanding some of these problems in greater detail may come from studies of the effect of ethanol and other alcohols on lipid membranes of different composition in the absence of proteins. It has been found that a small alcohol like ethanol preferentially localizes in the region of the glycerol backbone of the lipid bilayer. This leads to an increase of the lateral pressure in this region. Alcohols with longer hydrocarbon chains do this to a progressively lesser degree the longer the chain is. It is noteworthy that alcohols with increasing chain lengths become less potent anesthetics as the chain length gets longer. These observations can be rationalized within the picture of the lateral pressure profile in Fig. 8.2. Ethanol will shift the balance of the forces across the bilayer to build up a larger lateral pressure in the region of the glycerol backbone.

Once this is said, it would imply that the effect of ethanol could be reversed by adding special compounds that could shift the lateral pressure profile back toward the center of the bilayer. Cholesterol is a possibility. Indeed, it has been found that cholesterol acts to squeeze ethanol out of the bilayer, as demonstrated in Fig. 12.2b. Hence, cholesterol could be expected to reduce some effects of alcohol intoxication. It fact, mice subject to chronic treatment with ethanol do develop higher tolerance to alcohol, and their nerve membranes have been found to contain elevated levels of cholesterol. Similarly, one would expect that compounds that synergetically with ethanol act to shift the lateral pressures of the membrane toward the head-group region would enhance the effect of alcohol. Studies have shown that lipid bilayers incorporated with glycosphingolipids with saturated lipid chains bind more alcohol, possibly because the relative bulky head group renders the effective molecular shape more conical. This correlates with the physiological finding that mice fed with glycosphingolipids become sensitized to ethanol.

Robert Cantor has proposed a physical theory of general anesthesia that provides a mechanistic and thermodynamic understanding of these and related phenomena. The theory is in contrast to earlier lipid-based theories not based solely on empirical correlations between anesthetic potency and structural and thermodynamic parameters. The theory involves a mechanism by which drugs alter the lateral pressure profile of lipid bilayers. According to this mechanism, the magnitude of the shifts in the lateral pressure profile induced by alcohol and other anesthetics is enough, at clinical concentrations of the drugs, to induce sufficient variations on the stresses on integral membrane-bound channels to inhibit or potentiate conformational transitions, cf. Fig. 15.9.

17.3 Poking Holes in Membranes

There is a host of chemical substances that very aggressively bind to cell membranes and eventually destroy the cell by poking holes in its membrane. It can happen delicately by forming supramolecular channel aggregates in the lipid bilayer that disrupt the delicate ionic balance across the membrane, or it can happen more brutally by rupturing the membrane. Many of these substances are amphiphilic polypeptides that have a hydrophobic domain that shows affinity for membrane surfaces by the mechanism sketched in Fig. 13.1f. Obviously, peptides of this type could be toxic for the cells to whose membranes they bind, but may turn out to be harmless for other cell types. The big challenge is to figure out the molecular mechanisms by which the peptides work in order to find new and potent drugs that can help us fight microbial organisms. The demand for such new drugs is becoming pressing, since more microorganisms become resistant to conventional antibiotics such as penicillin, and multi-drug resistance is becoming a major problem for immunodepressed patients. The quest is to find a highly potent antibiotic that selectively binds to microbial membranes but not to the membranes of eukaryotic cells, in particular human red blood cells. Let us consider some polypeptides that are known to have lytic activity on membranes and therefore may be candidates for useful antibiotics.

The first candidate is the small amphiphilic polypeptide melittin from bee venom. Melittin is the most extensively studied lytic peptide. However, melittin does not show any specificity for any particular cell type and breaks down both prokaryotic and eukaryotic membranes. Hence, it would destroy the red blood cells if used intravenously as an antibiotic in humans. Another candidate is the polypeptide δ-lysin which is secreted from *Staphylococcus aureus*. δ-lysin does not kill bacteria but lyses red blood cells. Hence, none of these peptides would serve our purpose.

Another strategy would be to look for peptides that are a natural part of either plant or animal defense systems. This points to a large class of antimicrobial peptides. These peptides typically have 12–45 amino acids, they have some structural fold like α-helix or β-sheet, and they are basic. Some of these peptides are known to be potent bacteria killers. We now know part of the molecular mechanism by which they work on membranes. Two cases are particularly well studied. One is magainin 2, which has twenty-three amino acids. It is found in frog skin and appears to be part of the frog's natural defense against bacteria. The other is alamethicin, which has twenty amino acids and is found in fungi.

Both magainin 2 and alamethicin have an α-helical structure with a face that is hydrophobic. Hence, they will adsorb parallel to membrane surfaces with the hydrophobic face toward the membrane interior. The degree of penetration into the bilayer depends on the hydrophobicity of the peptide. Magainin 2 is less hydrophobic than alamethicin and does not penetrate very deeply. The adsorption leads to a thinning of the membrane, as indicated schematically in Fig. 13.1f. This, in turn, induces a positive curvature stress in the bilayer and would therefore counteract any propensity the target membrane may have for forming H_{II} lipid phases. For increasing peptide concentrations, this leads to a build up of strain that turns out to be released at a critical concentration where magainin leaves the parallel state and flips into a transmembrane orientation.

Once in the transmembrane configuration, the peptide tends to form supramolecular pore complexes, which break down the permeability barrier of the membrane. Alamethicin and magainin 2 form different types of pores, a so-called barrel-stave pore in the case of alamethicin and a so-called toroidal (wormhole) pore in the case of magainin 2, as illustrated in Fig. 17.1. In contrast to the barrel stave pore, the toroidal pore is lined by lipids at its bore. Not all antimicrobial peptides work in this way. Some appear to use a more detergent-like strategy (the so-called carpet mechanism) by dissolving a patch of the membrane, leaving a big hole behind.

Recently, an antimicrobial peptide has been found that is a thousand times more potent that magainin 2. This peptide is called nisin, and it works by a dual mechanism. The peptide is prenylated, i.e., it is linked to a long

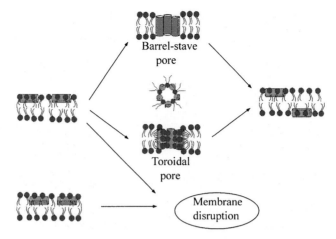

Fig. 17.1. Poking holes in membranes by antimicrobial peptides. Pore complexes formed by alamethicin (barrel-stave pore) and magainin 2 (toroidal pore)

lipid (so-called lipid II) of the strange type shown in Fig. 2.6d. The peptide is a very specific pore-former, and at the same time its lipid companion binds the material that the bacterium needs for its cell-wall synthesis.

17.4 Gramicidin – the Portable Hole

Gramicidin A is a small linear polypeptide with fifteen amino acids that forms a helical structure when embedded in membranes. This is illustrated in Fig. 17.2. The bore of the helix is 0.4 nm and precisely large enough to permit water molecules and small alkaline-like ions such as Na^+ to pass through. One gramicidin A molecule is like half a hole, which can span one monolayer of a bilayer membrane. When two gramicidin A molecules in opposed monolayers match up, they bind by hydrogen bonds and form a complete hole or channel through the bilayer, as illustrated in Fig. 17.3a.

The channel mediates ion leakage. This is why gramicidin A can be used as an agent to kill bacteria by destroying the ionic balance across their membranes. Gramicidin A was discovered in 1939 and became the first clinically useful topical antibiotic. It was used during World War II when penicillin was still scarce, but it is no longer in use.

The structure of the gramicidin A channel is known in great detail. Due to the hydrogen bonds, the gramicidin A channel is very robust, and it has a very well-defined function. It therefore lends itself to detailed biophysical studies. The activity of the channel is measured by monitoring the current flow by electrophysiological methods, and it is possible to measure the opening of single channels as well as the life time of each channel. This is illustrated in Fig. 17.3b and c. The current trace shows a baseline corresponding to no

17.4 Gramicidin – the Portable Hole   187

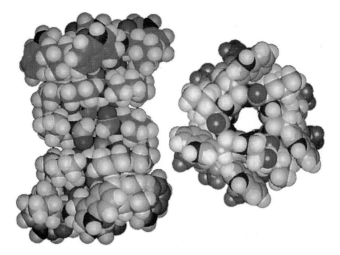

Fig. 17.2. Molecular model of a gramicidin A dimer seen from the side (*left*) and from the top (*right*) of the membrane in which it is embedded

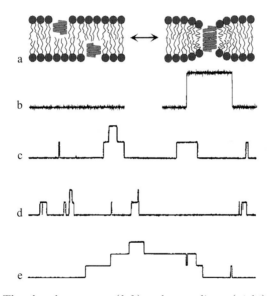

Fig. 17.3. (a) The closed monomer (*left*) and open dimer (*right*) configuration of gramicidin A in a lipid bilayer. (b) Electric current traces monitoring the activity of gramicidin A corresponding to the two states in (a). The electrical current is shown as a function of time. (c) Electric current trace for a specific membrane. Sometimes more than one channel is open. (d) Electric current trace for the same membrane as in (c) now incorporated with cholesterol, which tends to suppress the channel activity. (e) Electric current trace for the same membrane as in (c) now incorporated with lysolipids, which tend to enhance the channel activity

channel activity and then a burst of current corresponding to the opening of a single channel. Sometimes other channels open at the same time, leading to currents that are multipla of the single-channel current.

Although gramicidin A is no longer itself considered a useful drug, its well-defined mode of functioning as a model for transmembrane channel activity (cf. Sect. 15.3) can be exploited to study the pharmacological effects of other molecules, such as drugs, on membranes and membrane channels. To illustrate this point, Fig. 17.3d and e show how two different amphiphilic molecules, cholesterol and lysolipid, have dramatic and opposite effects on the channel activity. Since these compounds do not bind chemically to gramicidin A, their effects on the channel activity have to be mediated by the lipids. The opposite modes of action of these two compounds can be rationalized by their differential effect on the lipid-peptide boundary. Cholesterol promotes negative curvature, whereas lysolipid promotes positive curvature, cf. Fig. 4.6.

18 Lipid Eaters

18.1 Enzymes That Break Down Lipids in Crowds

There is a constant turnover of lipids when a cell or an organism performs its life functions. Lipids have to be molecularly remodelled in order to meet the needs of a particular cell or tissue type, lipids have to be broken down to fatty acids and mono-acylglycerols in order to be able to be transported across membranes, and lipids have to be exported to where they are needed either as fuel, structure builders, or signal molecules. There is a host of catalysts in the form of enzymes that help make these processes possible.

Enzymes that facilitate degradation of lipids and fats are called lipases, and the process of degradation is referred to as *lipolysis*. Lipolysis requires only little energy, and it takes place both within cells, as well as outside cells such as in the blood stream and in the gut. *Phospholipases* constitute a ubiquitous class of special enzymes that selectively can break down phospholipids. These enzymes are widespread in nature. Some of them are digestive enzymes that are found in, e.g., venoms, bacterial secretions, and digestive fluids of animals. Others are used in the remodelling of membranes such as neural membranes, as described in Sect. 16.3, or in forming the permeability barrier of the skin, as described in Sect. 19.1. Still others are involved in regulatory functions and cell signalling cascades, often in association with membranes, by producing special lipids like di-acylglycerol, phosphatidic acid, and ceramide, as described in Sect. 19.3. As an example, our tear fluid contains a secretory phospholipase that attacks bacterial membranes and hence functions as part of the body's defense systems.

The phospholipases are divided into several families depending on where they can cut a lipid molecule into two or more pieces. As illustrated in Fig. 18.1, phospholipase A can cut off a fatty-acid chain at the glycerol backbone. Depending on which chain is cut off, one speaks about phospholipase A_1 and phospholipase A_2. The result of the cleavage is a fatty acid and a lysolipid. Phospholipase B can cut off both fatty-acid chains of a di-acyl glycero-phospholipid. Phospholipase C can cut off the head group, thereby producing di-acylglycerol. Phospholipase D can cleave the base group off the polar head, leading to phosphatidic acid. In the case of sphingolipids, there is an enzyme called *sphingomyelinase*, which, similar to the action of phospholipase C on phospholipids, can cut off the head group of sphingomyelin,

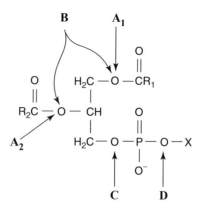

Fig. 18.1. Action of lipases, enzymes that can break down lipids into two or more pieces. (**a**) Phospholipase A_1 can cut off a fatty-acid chain of a di-acyl phospholipid in the *sn*-1 position. (**b**) Phospholipase A_2 can cut off a fatty-acid chain of a di-acyl phospholipid in the *sn*-2 position. (**c**) Phospholipase B can cut off both fatty-acid chains of a di-acyl phospholipid. (**d**) Phospholipase C can cut off the head group of a phospholipid. (**e**) Phospholipase D can cut off the base part of the head group of a phospholipid. R_1 and R_2 are the two fatty-acid chains, and X is a variable part of the head group

thereby producing ceramide. There are several types of sphingomyelinase, some are of major importance for the formation of the human skin, as discussed in Sect. 19.1, and others for programming cells to commit suicide, as described in Sect. 19.4.

Whereas plants can synthesize their own lipids where needed, vertebrates, like humans, need to get lipids via the diet. In order to utilize the lipids in the foodstuff, or to take advantage of lipids that have already been stored in fat depots like adipose tissue, animals have to perform a number of tricks. Let us follow the fate of fats as they are consumed by a human being. If not already in particulate form, like milk fat in the form of micelles, the fats have to be mechanically churned and turned into tiny globules. This happens in the stomach. Since these fat globules are insoluble in water, they have to be emulsified before enzymes can start working on them. The emulsification is facilitated by the bile salts that are produced in the gall bladder and injected into the intestine. The bile salts, which are compounds similar to cholesterol, are interfacially active molecules that make fats soluble in water, as discussed in Sect. 3.3. Only at this stage can the enzymes start their work.

A requirement for the action of the enzymes is that the lipids appear in crowds such that they present themselves to the attacking enzymes in the form of interfaces. An emulsion is therefore susceptible to enzyme attack since it is basically a bunch of interfaces, e.g., in the form of micelles, monolayers, or bilayers, as illustrated in Figs. 3.3 and 3.4. The enzymes are water soluble and therefore have to attack the fat from the watery side. Some mammals, including humans, produce a gastric lipase that already starts working on the emulsion in the stomach. Otherwise, the pancreas produces digestive lipases and phospholipases that act in the small intestine. The lipases turn tri-acylglycerol into fatty acids, and the phospholipases turn phospholipids into fatty acids and lysolipids.

The breakdown is faster for lipids with short- and intermediate-length fatty acids such as milk fat, whereas it is much slower for fats with long fatty acids such as fish oil. Hence, essential fatty acids take much longer to be released in the digestive process. Infants of some mammals like humans also have some digestive enzymes in the saliva in the mouth. Hence, the breakdown of, e.g., mother's milk, is initiated in the mouth. In fact, mother's milk itself contains lipases that facilitate this early process. In the form of fatty acids and mono-acylglycerols, the fats can be adsorbed and transported across the cell membranes in the gut and make their way into the blood. Although this transport seems to be facilitated by special proteins that carry the fats across the membrane, there are still major unresolved questions as to which digestive products cross the intestinal barriers. Once in the blood, the fatty acids are transported by a number of proteins that, together with the lipids, form lipoproteins, as mentioned in Sect. 2.1.

In order to illustrate the association of an active lipase with a crowd of lipid molecules, Fig. 18.2 shows a phospholipase A_2 molecule bound to a monolayer of phospholipids. The phospholipase molecule is partly penetrating the head-group region of the monolayer whereby a target lipid molecule becomes exposed to the active site of the enzyme. This is the necessary event

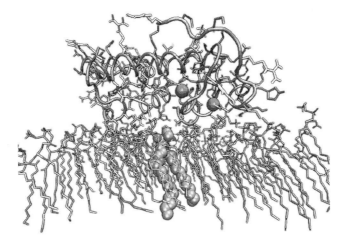

Fig. 18.2. Molecular representation of a phospholipase A_2 molecule bound at the watery side of a phospholipid monolayer. The picture is obtained from an atomistic model of the system studied by Molecular Dynamics calculations. A single phospholipid molecule prone to attach is enhanced. This molecule is near the active site of the enzyme where the hydrolytic cleavage of the lipid takes place

that precedes the catalytic step. The activation of the enzyme and the actual hydrolytic process are extremely sensitive to the structure and quality of the lipid bilayer, in particular, the cooperative behavior of the lipid assembly, as described in Sect. 12.3. A typical turnover rate of the enzyme is several hundred lipid molecules per second.

It is interesting to note that the hydrolysis products of phospholipase A_2, i.e., fatty acid and lysolipid, have a propensity for forming non-lamellar lipid structures, as discussed in Chap. 4. This may provide biological cell membranes with a mechanism of regulation via changes in the lateral curvature stress field which follows from the enzymatic cleavage of the lipids.

18.2 Watching Enzymes at Work

It is possible to observe directly the action of phospholipases that are in the process of breaking down a lipid monolayer or bilayer using atomic force microscopy, as described in Sects. 10.3 and 11.2. Figure 18.3a shows a series of pictures of a small portion of a solid-supported lipid bilayer that is being eaten by a phospholipase. It is possible to follow the process in real time and investigate where the bilayer is most susceptible to degradation. It turns out that the enzyme is most active where the bilayer has defects. This process is self-enhancing, since the hydrolysis products, before leaving the supported layer, will themselves play the role of defects and hence enhance the enzyme action. Similarly, pre-existing defects like holes in the bilayer are sites that activate the enzyme. It is not yet possible however to observe the action of a single enzyme by this method, although single enzymes are often seen on the images.

A second illustration of phospholipases at work is given in Fig. 18.3b, which shows the time evolution of the action of phospholipase A_2 on a phospholipid bilayer in water observed by atomic force microscopy. In this case, the lipid bilayer is a mixture of phospholipids with different fatty-acid chain lengths, DMPC and DSPC. The mixture is studied in the region of phase separation where domains of liquid and solid lipids coexist, cf. the phase diagram in Fig. 9.7. The pictures show that the short-chain lipid species DMPC, which predominantly makes up the liquid domains, is the favorite food of the lipid-eating enzyme. Hence, as the enzymatic process proceeds, the liquid domains of the bilayer are eroded away and exposed as holes in the membrane.

An ultimate goal would be to be able to control and monitor the action of a single or a few active enzymes at the tip of an atomic force microscopy using techniques illustrated in Fig. 13.2. This is not yet possible. However, it is feasible to attach enzymes to a small latex bead whose position can be manipulated by micro-pipettes. It is then possible to follow, on the scale that is accessible by light microscopy, the action of the enzymes on giant lipid vesicles. This is illustrated in Fig. 18.4, which shows an experiment using sphingomyelinase that is immobilized on a latex bead. The bead is

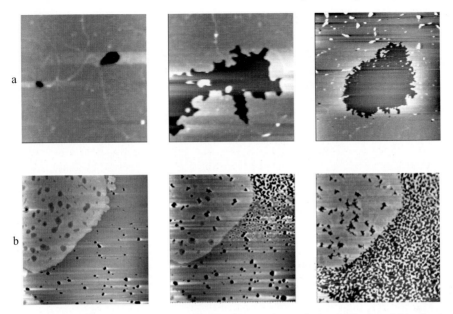

Fig. 18.3. (a) A solid-supported DPPC bilayer that is being hydrolyzed by phospholipase A_2. The image to the *left* (2 μm × 2 μm) shows the bilayer before adding the enzyme. Two holes in the bilayer can be seen. The middle image (2 μm × 2 μm) shows the same frame after the enzyme has been added. The image to the *right* (6 μm × 6 μm) shows how one of the initial holes has been enlarged after about 20 min due to the action of the enzyme. (b) Time evolution of a phospholipid bilayer in water under the action of phospholipase A_2. The bilayer is composed of a 1:1 binary mixture of DMPC and DSPC in the solid-liquid phase separation region, cf. the phase diagram in Fig. 9.7. The light regions are the solid phase consisting predominantly of DSPC molecules, and the light-dark regions at early times are the liquid phase consisting predominantly of DMPC molecules. As time lapses, the enzyme predominantly hydrolyzes the DMPC patches, which then turn darker as the hydrolysis products leave the bilayer. The image sizes are 5 μm × 5 μm

then approached to the surface of a giant vesicle containing sphingolipids. The formation of ceramides can then be followed by fluorescence techniques.

18.3 Lipids Going Rancid

It is well-known that lipids and oils can be broken down by burning, i.e., oxidation, as it happens when wax and oil are burned in candles and lamps. In this case, the released chemical energy is readily transformed into heat and light. When lipids are used as an energy source in living systems, the oxidation has to be much slower and well controlled in order to be useful in fueling other chemical reactions in the cells. The controlled oxidation can

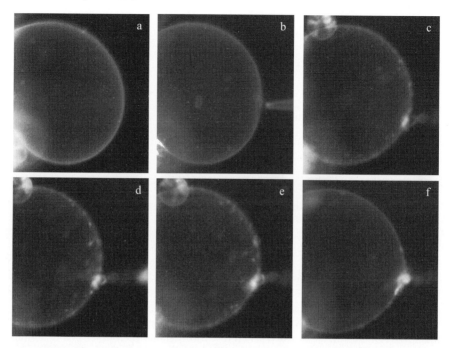

Fig. 18.4. Micro-pipette techniques are used to turn sphingolipids into ceramide by approaching a latex bead covered with sphingomyelinase to the surface of a giant vesicle containing sphingomyelin. The patches of formed ceramide are lightning up due to fluorescence. Time lapses from frame *a* to *f*

only occur after the lipids and the tri-acylglycerols have been hydrolyzed by enzymes into fatty acids and glycerol. Sugars contain more oxygen than fats and, therefore, produce less energy than fat when fuelling the body. However, sugars are more readily burned, which leads to faster release of energy.

Lipids in food are also subject to spontaneous breakdown. Fatty fish like salmon and mackerel become smelly, and fatty nuts like walnuts and hazelnuts go rancid. This breakdown is due to the oxidation of the lipids and is different from the effects of the deterioration of proteins that occurs when the food stuff is contaminated by bacteria and fungi. The consequences of lipid oxidation is lower nutritional value, and, in some cases, the oxidation products are toxic. Unsaturated and polyunsaturated lipids are most susceptible to oxidation, and the oxidation is facilitated by light and high temperature. The oxidation can be prevented by so-called anti-oxidants, e.g., ascorbic acid and vitamin E.

A particularly troublesome type of oxidation is so-called peroxidation, which involves much more reactive oxygen-containing molecules than ordinary oxygen, e.g., hydrogen peroxide. These molecules are called *reactive oxygen species* (ROS). ROS can be formed as byproducts of the natural oxidative processes in the mitochondria of animals and in the chloroplasts of green plants. Tobacco smoking and radiation are also known to increase the level of ROS. When ROS are produced in uncontrollably large amounts, the organism is brought into a state of oxidative stress. Oxidative stress can influence signal pathways, change enzyme activities, and cause damage to proteins and DNA, which may lead to mutations. ROS are known to be involved in common and serious diseases such as atherosclerosis, cataract, Alzheimer's disease, and colon cancer. Undoubtedly, ROS also play a role in the general process of aging.

The peroxidation process of unsaturated and polyunsaturated lipids starts by removal, via ROS, of a hydrogen atom from a methylene CH_2-group of a fatty-acid chain or the ring structure of a steroid-like cholesterol, leading to the formation of a lipid reactive oxidative species. The oxidation can propagate, as a kind of chain reaction, along the fatty-acid chain by migration of double bonds, by further reaction with oxygen, or by converting a neighboring fatty-acid chain into an ROS. The process can be terminated by anti-oxidants or by combining two ROS. In order to fight the production of ROS, the human body has developed a series of anti-oxidant defense mechanisms involving chemical substances that can donate an electron to the ROS species, e.g., intracellular superoxide dismutase, catalase, and glutathione peroxidase. The hormone melatonin is also believed to be an important anti-oxidant in the body.

19 Powerful and Strange Lipids at Work

19.1 The Impermeable Barrier – Lipids in the Skin

The skin is our largest organ and anatomically one of the most heterogeneous. Its presence and function are usually taken for granted, and few of us wonder why we are not dissolved and flushed down the drain when we take a shower. We tend to think of the skin as an organ that is inferior to other organs like the heart, the brain, and the liver. However, the skin is a remarkable organ whose unique barrier properties are largely determined by lipids. The skin provides us with the necessary protection against a hostile environment with hazardous chemicals, radiation, microbial attacks, as well as mechanical stress. It is also instrumental for retaining the precious materials of our inside, not least water. Lipids like cholesterol, fatty acids, ceramides, and sphingolipids are essential for providing the skin with these unusual barrier properties.

The main permeability barrier of the skin is the outermost horny part of the skin, the so-called *stratum corneum*, as illustrated in Fig. 19.1. The stratum corneum is typically $10\,\mu m$ thick but can be much thicker at high-friction surfaces, like the soles of our feet and the palms of our hands. Stratum corneum is a dead layer of tissue, resting on the viable epidermis. The epidermis is a live tissue from which the stratum corneum is grown by a process that determines the particular composite structure of the permeability barrier. This detailed structure of the composite is complex and to some extent unknown. Several models have been proposed. One of them compares the composite with a wall of bricks and mortar where the mortar is a bunch of lipid bilayers and the bricks are flaccid and dead protein-containing cells, the so-called corneocytes. The stratum corneum contains a certain amount of water, but the detailed distribution of the water remains at present unknown.

Although there are no blood cells in the epidermis, it functions in a way analogous to that of blood. Blood cells are the mature and differentiated stage of cells produced from the so-called stem cells that reside in the spinal cord. The mature red blood cells function about 120 days and are then destroyed and recycled. Similarly, the corneocytes, which in the present analogy correspond to the mature blood cells, are mature forms of cells derived from stem cells or so-called keratinocytes residing at the bottom of the epidermis, next to the dermis. The dermis has a vascular system that feeds these cells. The maturing corneocytes move up toward the top of the epidermis

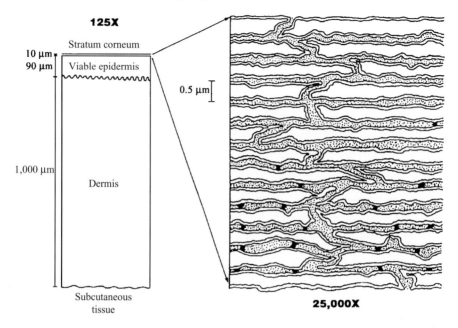

Fig. 19.1. Schematic illustration of a cross section of the human skin highlighting the composite structure of the stratum corneum. The overall structure can be compared to that of a wall with solid bricks of dead corneocyte cells and with mortar made of stacks of lipid bilayers

over a period of two weeks and eventually get lost at the surface of the skin. The corneocytes are filled with proteins called keratins that are arranged in a tight fibrous structure. Upon arrival in the stratum corneum, the corneocytes become glued together of intercellular connections and eventually form the bricks of the wall.

In parallel to this process, the keratinocytes, during the maturing process, have been proposed to nurture small organelles in their interior, so-called lamellar bodies that are loaded with lipids and lipases that can modify lipids. The lamellar bodies are expelled from the keratinocytes by exocytosis and their lipid contents eventually become the mortar that fills the intercellular space between the corneocyte bricks of the stratum corneum, as illustrated in Fig. 19.1. Recently, the Swedish chemist Lars Norlén suggested an alternative model of the formation of the stratum corneum that involves a continuous unfolding transition from the *trans*–Golgi membrane network of the keratinocytes, which is of cubic structure, to the lamellar morphology of the lipid bilayers of the stratum corneum mortar. This process is illustrated in Fig. 19.2. Within this picture, the lamellar bodies are local regions of the folded structure that contains densely packed lipids in a solid-like phase. Concomitant with the unfolding process, the lamellar lipid sheets become more crystalline.

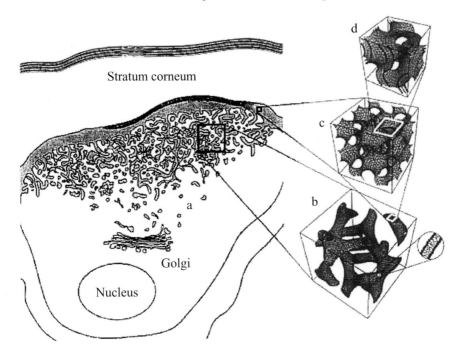

Fig. 19.2. Schematic illustration of the continuous unfolding process (**b–d**) of the cubic membranes from the *trans*–Golgi network (**a**) into the lamellar structures of the lipid membranes of the stratum corneum part of the skin

There are several reasons for this solidification process. One being the decrease of water content and therefore the level of lipid hydration as the surface of the skin is approached. The other being a change in lipid composition from the epidermis toward the stratum corneum. The epidermis contains predominantly mono- and di-unsaturated phospholipids. Although the lipid composition is difficult to analyze and varies through the stratum corneum, the main components appear to be an equimolar mixture of free fatty acids, ceramides, and cholesterol. There is hardly any phospholipid. The free fatty acids and the ceramides are formed, as part of the maturing process, from the lamellar bodies' contents of phospholipids and sphingolipids by enzymatic processes involving, e.g., sphingomyelinase and phospholipases. As described in Chap. 18, sphingomyelinase turns sphingolipids into ceramides. Hence, lipid metabolism is absolutely essential for forming the peculiar structure of the skin. It is a striking observation that the chains of the free fatty acids and the ceramides in stratum corneum are very long. Ninety percent of the chains of the free fatty acids are 16:0, 18:0, 18:1, and 18:2, and 75% of the chains of the ceramides are 18:2, 24:0, and 26:0. Since the long-chain lipids have high transition temperatures, this implies that the bilayers are predominantly in a solid phase.

Ninety percent of the lipids in the skin are localized in the mortar region. The lipid mortar hence becomes the only continuous element of the complex composite that makes up the stratum corneum and thus constitutes the seminal permeability barrier of the skin. The lipids and free fatty acids in the stratum corneum are organized in multi-lamellar arrays parallel to the skin surface, as shown in Fig. 19.1. In fact, it has been suggested that the stratum corneum lipids form a mosaic of crystalline domains glued together of a small fraction of lipids in the fluid phase. Whether this is the case or the bilayers are in a single solid phase, as suggested by Lars Norlén, the lamellar lipid arrangement forms a perfect waterproof layer and an almost ideal barrier for both strongly polar and strongly nonpolar molecules.

The skin is an outstanding example of nature's way of solving a complicated barrier problem by using powerful lipids that form soft matter with unique physical properties. Trespassing this barrier is not easy. This poses an outstanding problem in the pharmaceutical sciences where the dermal route for applying drugs in many cases may be desirable. As a prominent example, it would be an enormous advantage for diabetic patients, rather than administering insulin via injection into the blood, to apply a spray with insulin at the nasal skin. This is not yet possible.

Recent research has shown that the dermal route may be opened by using pulsed electric fields at the skin, or by applying certain permeability enhancers, some of which are lipids. The electric fields are believed to form transient pores through the stratum corneum, whereas the enhancers seem to change the phase properties of the lipid bilayers in stratum corneum. These phase properties can also be altered by changing the gradient in the water chemical potential across the skin, from the dry surface to the moist blood-containing dermis, e.g., by applying drugs at moist skin. The moisture increases the hydration of the lipid bilayers and enhances the partitioning of the applied drug into the skin. As an example, nicotine diffusion across the skin is facilitated by being applied in the moist. In Sect. 20.2, we shall describe an example of how certain soft liposomes, so-called transfersomes, appear to enjoy transport across the skin by being drawn through the water chemical potential gradient. Transfersomes can be perceived as soft bags that can carry drugs across the skin by squeezing themselves through the lipid mortar of the stratum corneum.

19.2 Surviving at Deep Sea and in Hot Springs

As human beings we consider normal living conditions to imply an environment of certain agreeable physical and chemical properties. The temperature has to be in the range of our body temperature, the pressure has to be around 1 atmosphere, and the chemical conditions should not be to be too salty, too acidic, or too basic. Still, there are other organisms for whom these conditions would not support their forms of life. These organisms are mostly prokaryotes, eubacteria or archaebacteria, who have adapted to what we would call extreme conditions. In our eyes, they live a hard life characterized, e.g., by very high or very low temperatures, very high pressure, or extreme conditions in terms of saltiness, acidity, or high levels of chemicals that would be poisonous to us. We call these organisms *extremophiles*. Obviously, this is a relative term. Extremophiles could, with good right, consider humans and other animals as extremophiles, since we require in our habitats large amounts of an extremely reactive chemical species, oxygen, which is poisonous for many prokaryotes.

In order to survive under extreme conditions, the membranes and the proteins of the extremophiles have to be rather different from those of eukaryotes and most other prokaryotes. In particular, nature has evolved special "strange" lipids of a particular molecular structure and effective shape in order to provide the membranes of the extremophiles with proper physical conditions to support their biological activity. The maintenance of the liquid state of the membranes appears to be of particular importance. Although very little is at present known about this, it would be expected that homeostatic control of the curvature stress and the lateral pressure profile of the lipid bilayers is equally important.

Eubacteria that prefer habitats of elevated temperatures, in the range from $50°$ to $113°$, are called thermophiles. Such high temperatures may occur around deep-sea volcanic vents. The strategies used by these organisms to preserve the stability of lamellar lipid membranes and maintain the proper fluidity and mechanic coherence involve, e.g., use of very large sugar head groups that are tightly bound by hydrogen bonds. Some organisms also benefit from lipids with three rather than two fatty-acid chains that allow for stronger coherence of the hydrophobic part of the bilayer. Moreover, thermophiles often use ether lipids that tolerate higher temperatures than the ester lipids used by, e.g., eukaryotes.

Whereas thermophiles like it hot, the so-called psychrophiles have adapted to the harsh cold conditions that prevail, e.g., in the Antarctic sea deep below the ice. At the elevated pressures in these waters, the temperature drops below $0°C$. In order to maintain the liquid character of the membranes at these low temperatures, where many eukaryotic membrane lipids would simply solidify, the psychrophiles use special lipids with very low melting points. Examples include lipids with branched chains, lipids with short chains, and lipids with a number of double bonds.

A certain class of bacteria has evolved to sustain the extremely high pressures that exist in deep oceans. Some of these bacteria, which are called piezophiles, live 10,000 meters below the surface of the sea and, therefore, experience pressures in the range of 1,000 atmospheres. At the same time, the temperatures may be in the range of $-0.5°C$ to $113°C$. A typical habitat would be cold waters, around $2°C$. A hydrostatic pressure of 100 atmospheres corresponds to an elevation of the melting point of lipids of the order of

2–8°C. Hence, both the high pressures and the cold water act to solidify lipid membranes. The reason why such high pressures only lead to a rather small change in melting temperature is that, as described in Sect. 9.2, the chain-melting transition in lipids is associated only with a modest change in volume.

To counteract this physico-chemical phenomenon, the piezophile bacteria incorporate large amounts of polyunsaturated fatty-acid chains in their lipids, e.g., 20:5 and 22:6. The fact that polyunsaturated lipids otherwise are almost absent in the bacterial world corroborates the suggestion that the piezophiles keep their membranes in the physiological liquid state by taking advantage of the disorder that the polyunsaturated lipids impart to their membranes. It is interesting to note that bacteria grown under varying pressure conditions compensate the pressure effects on their membranes by varying the ratio between saturated and unsaturated fatty-acid chains in the lipids. In Sect. 4.4, we saw a similar homeostatic principle operative in the case of another microorganism, *Acholeplasma laidlawii*, which under varying growth conditions maintained the curvature stress field in its membrane by varying the ratio of lipids with small and large head groups.

Some of the most bizarre and fascinating ways of developing lipid-based strategies to survive under extreme conditions are found in the kingdom of the archaebacteria. This kingdom, which was identified as a separate kingdom of life only a quarter of a century ago by the American microbiologist Carl Woese, encompasses extremophiles that live in deep-ocean hot volcanic vents, in very salty water, in the acid guts of animals, and under harsh chemical conditions with high levels of, e.g., methane or sulphur.

The lipids constituting the membranes of archaebacteria are very different from those used by eubacteria and eukaryotes. First of all, the chemistry of their fatty-acid chains is based on poly-isoprene, which forms so-called phytanyl (or isopranyl) chains. These chains are fully saturated and have a methyl group sticking out from the chain at every fourth carbon atom, as shown in Fig. 2.6b. The protruding methyl groups presumably serve the same purpose as double bonds in eukaryotic membranes: They keep the lipid melting transition temperatures sufficiently low even at high pressure. Moreover, the phytanyl chains are always connected to the glycerol backbone by ether bonds rather than ester bonds. In some cases, the glycerol backbone is replaced by another longer alcohol, which leads to a much larger head group. Often the isoprene units are cyclized to form rings of five carbon atoms on the chain. Finally, the ends of the two phytanyl chains of a di-ether lipid of this type can be chemically linked to the corresponding ends of another lipid of the same type, forming a so-called tetra-ether lipid, as shown in Fig. 2.6c. This type of lipid is called a *bolalipid* since it has a head group in both ends and can span the entire membrane if the phytanyl chains are long enough. This is illustrated schematically in Fig. 19.3.

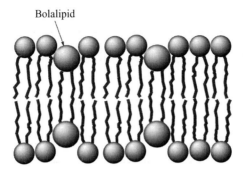

Fig. 19.3. Schematic illustration of a lipid membrane of archaebacteria with bipolar bolalipids that span the membrane

As an example, the thermophile *Thermoplasma acidophilum* has forty carbon atoms in its transmembrane tetra-ether lipids. This leads to a hydrophobic membrane thickness in the desirable range of 2–3 nm. Very little is known about the physical properties of membranes and lipid bilayers made of these strange lipids. However, it can be surmised that the isoprenic character of the hydrocarbon chains, in particular, in the case of cyclization along the chain, provide for chain disorder and liquid membranes even at high pressures and low temperatures. Moreover, the bolalipids are expected to provide for additional mechanical stability and appropriate membrane permeability properties under harsh chemical conditions.

19.3 Lipids As Messengers

The functioning of individual cells as well as assemblies of cells relies on a host of communication systems and signal pathways. Some types of communication act over long distances, others involve the transmission of signals over short distances or may even require close contact of the involved cells. In all cases, specific signal molecules are involved, e.g., peptides, proteins, nucleotides, steroids, gases, as well as fatty acids, lipids and their derivatives. For example, hormones like adrenaline and insulin are secreted from glands into the blood stream and are thereby carried to distant receiving cells. In general, the communication pathway involves a number of steps, a so-called signalling cascade. Normally, the signal has to be converted from one form into another in order to pass a barrier, e.g., a membrane or the space between neighboring cells. This type of signal transduction often involves receptor proteins that can recognize the signal and convert it to another signal. The signalling cascades can be very complex with many steps and intermediates as well as feedback loops.

Some of these receptor proteins are bound to membranes like the integral membrane proteins we discussed in Sect. 13.3. The membrane-bound receptors can transmit signals carried by molecules that themselves are too large to directly transmit their information into the cell by permeating through the membrane. In some cases, the membrane receptors, after binding the signal molecule, interact with an enzyme inside the cell. This enzyme then becomes activated and produces a new signal molecule, a so-called second messenger, the "first" messenger referring to the signal molecule that first bound to the receptor on the outside of the cell.

This is the point where lipids and lipid derivatives come in as powerful second messengers that are produced by lipid-modifying enzymes that are stimulated upon binding to membrane receptors. It has in recent years become increasingly clear that the membrane is a rich reservoir for recruiting second messenger molecules in addition to being a target for and mediator of cell signalling. Since the functioning of many lipid-modifying enzymes, e.g., phospholipases, is very sensitive to the physical state of the membrane, the signalling cascades can be triggered by alterations in the physical properties of the membrane. We shall illustrate this briefly with a few examples and in Sect. 19.4 describe in more detail how ceramides act as lipid messengers of death.

One example involves protein kinase C, which we also discussed in Sect. 15.2. Upon stimulation of the cell by, e.g., neurotransmitters, hormones, and growth factors, the transmembrane receptor activates a phospholipase C. The activated enzyme hydrolyzes inositol phospholipids at the inner leaflet of the plasma membrane. This leads to di-acylglycerol (cf. Fig. 2.2d), which remains in the membrane because it is a lipid. The cleaved-off inositol-triphosphate head group dissolves in the cytosol and diffuses to the endoplasmic reticulum, where it leads to a release of Ca^{2+}. The calcium ions subsequently activate protein kinase C to bind to the di-acylglycerol remaining in the membrane. Upon binding, protein kinase C phosphorylates other proteins that influence cell growth as well as gene expression. In this complex signalling cascade, the lipid di-acylglycerol is a key messenger. As suggested in Sect. 15.2, it is likely that it is the propensity of di-acylglycerol for forming H_{II} phases that facilitates the binding of protein kinase C to the membrane.

Another example involves phospholipase A_2, which can cleave off arachidonic acid from phospholipids in plasma membranes. Arachidonic acid is the precursor of a very important class of hormones that function by binding to cell-surface receptors. These hormones are the so-called *eicosanoids*, e.g., prostaglandins, thromboxanes, and leukotrienes. The eicosanoids are involved in signal cascades that regulate inflammatory responses, blood flow, blood clotting, as well as modulate the contraction of smooth muscles. One of the intermediate enzymes that lead to the formation of prostaglandins and thromboxanes can be suppressed by aspirin and other anti-inflammatory drugs. In this way, aspirin reduces pain and inflammation as well as reduces the formation of blood clots. The eicosanoids work in extremely low concentrations, often one part in a billion, and their lifetime is restricted to seconds after their synthesis.

A number of other lipid-based signalling pathways are currently being explored and discovered. Amusing examples include certain brain lipids that induce sleep and lipid messenger molecules that help control pain. It has also recently been found that lipids may control signalling peptides involved in the motility of slime molds.

19.4 Lipids As a Matter of Death

A natural part of a cell's life is that it has to come to an end, and the cell must die. There are basically two ways for a cell in a multicellular organism to close its life. The first one is by so-called necrosis whereby the various functions of the cell stop because the tissue or organ it belongs to has stopped functioning. The second way is by cell suicide, also called programmed cell death or *apoptosis*. Apoptosis is a very tightly regulated process that can involve certain lipid messengers. These lipid messengers of death are ceramides, shown in Fig. 2.4a. Apoptosis is a crucial and necessary part of the life of a multicellular organism like a human being. When the natural mechanism of apoptosis for some reason is disturbed and the cells keep on living and multiplying, it leads to a diseased condition like cancer.

Necrosis is a rather uncontrolled process whereby the cell swells and its internal enzymes are released into the environment, thereby harming the neighboring cells. In contrast, apoptosis involves a controlled shrinking of the cell; the cell membrane remains intact, while the components of the cell are broken down and the resulting molecular constituents are transported away to be used in other cells. Apoptosis is therefore important for organogenesis, remodelling of tissue, removal of cells, as well as maintenance of the skin permeability barrier, as discussed in Sect. 19.1. Examples of cells undergoing necrosis and apoptosis are shown in Fig. 19.4a and b, respectively. During apoptosis, the cell plasma membrane loses its characteristic lipid asymmetry and its surface morphology changes by blebbing.

The details of the signalling cascades involving ceramides remain controversial. It adds to the complexity of the problem that ceramide can act as a second messenger in several other cellular processes in addition to apoptosis such as cell differentiation and growth suppression. Ceramides can be formed in the cell by *de novo* synthesis or by hydrolysis of sphingomyelin, a process that is catalyzed by several different kinds of sphingomyelinase. A stimulation of sphingomyelinase is therefore often involved in the ceramide signalling cascade. Later in the process, the choline head group and the ceramide are recombined into sphingomyelin whose original level is restored. Ceramides come in many different varieties with chain lengths ranging from

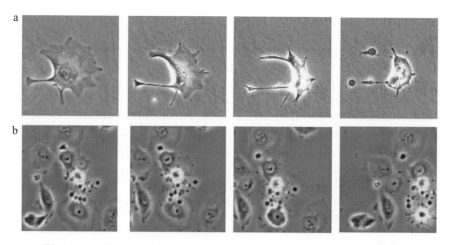

Fig. 19.4. Cells in the process of dying. (a) Necrosis. (b) Apoptosis

two to forty carbon atoms, but it seems as though palmitic chains are required for apoptosis.

Whereas cells under normal conditions contain very little ceramide, the ceramide content is increased up to about 10% of the lipid content upon apoptosis. The precursor sphingomyelin for the ceramide production is found predominantly in the outer leaflet of the plasma membrane. Due to its strong hydrophobicity, the produced ceramide will stay in the membrane in which it is formed. Since ceramide is a lipid molecule that has a very small head group, it is expected that an increase of the amount of ceramide in a cell membrane will lead to a significant increase in the membrane's propensity for forming non-lamellar phases, specifically H_{II} structures, cf. Sect. 4.3. In fact, Paavo Kinnunen has advocated the viewpoint that the effects of ceramide on the physical properties of the cell membranes are related to the molecular mechanisms behind apoptosis.

Model studies using giant vesicles containing sphingolipids have shown that upon action of sphingomyelinase, the produced ceramides aggregate in the membrane and form micro-domains, possibly of a solid or crystalline character. Subsequently, the micro-domains form small vesicles that are shed from the opposite side of the lipid membrane where the sphingomyelinase has been applied. This is illustrated in Fig. 19.5a. The vesicle blebbing is hence a vectorial process induced by the enzyme. If the sphingomyelinase is injected into the liposome, the blebbing is found to take place on the outside of the liposome. These dramatic events following the transformation of sphingomyelin into ceramide can be understood by noting that ceramide has a propensity for forming curved membranes, as illustrated in Fig. 19.5b. The smaller head group of the ceramide leads to a tighter packing of the fatty-acid chains in the membrane, which is the cause for the formation of solid micro-domains.

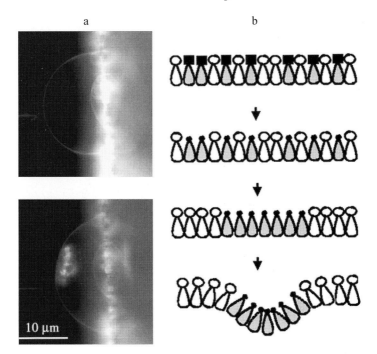

Fig. 19.5. (a) Sphingomyelinase added to the outside side of a liposomal membrane containing phospholipids and sphingolipids leads to blebbing of small vesicles on the inside of the liposome. (b) Schematic illustration of the molecular mechanism of vesicle budding induced by the formation of solid ceramide domains in the outer monolayer leaflet of the membrane. Due to the small head group of ceramide, these domains have a propensity for forming H_{II} structures

In fact, ceramide-containing membranes have a much higher melting point than membranes made of sphingomyelin. The observation of the dramatic changes of membrane morphology in these model studies suggests that the observed membrane blebbing processes, so characteristic of cells undergoing apoptosis, may be caused by a sphingomyelinase-controlled mechanism probably stimulated by signalling proteins that are localized in sphingolipid-rich domains or rafts in the plasma membrane, cf. Sect. 11.3.

A deeper understanding of the molecular mechanisms underlying apoptosis holds a promise for developing novel therapies for dealing with fatal diseases. Obviously, a controlled stimulation of the signalling behind apoptosis using appropriate anti-cancer drugs may suppress the proliferation of cancer cells. Similarly, a suppression of the signal pathways to apoptosis will be important for dealing with certain degenerative diseases that involve hyperactive apoptosis and massive undesired cell death.

20 Survival by Lipids

20.1 Lipids for Smart Nanotechnology

We often tacitly assume that useful devices have to be made of materials that are hard and solid in order to be tough, durable, and functionally reliable. A house, a car, and a computer are excellent examples. Although lipids and other soft materials can be used as templates for producing certain types of hard materials, we would not be using the full potential of lipid-based materials if we did not exploit the fact that they are based on self-assembly principles, they are designed to operate as soft materials, and their properties are optimized to function on a small scale. Functional macromolecules (proteins, enzymes, polynucleotides) as well as molecular assemblies (membranes, fibers, molecular actuators, and motors) are all designed by evolution to function optionally on the nanometer scale.

Insights into nature's design principles for soft natural materials are lessons for novel technology and promising applications within tomorrow's biomedicine and biotechnology, e.g., rational molecular drug design, smart and intelligent liposome-based drug delivery and cancer therapy, genedelivery and generepair, functionalized surfaces and sensors for biological and chemical recognition, biologically inspired computer technology, nano-wires, -flasks, and -networks, nanoscale structured depots for slow drug release, as well as fluid-flow nanoscale channels, pumps, and valves.

A particular perspective arises in interfacing hard and soft lipid matter on the nanoscale, e.g., with respect to combining biochemical processes in soft matter with signal processing in hard templates in order to produce biomimetic and self-healing materials with particular surface characteristics, sensors and nano-machines, as well as coupling of macromolecules and cells to hard templates. This type of interfacing provides a bridge between living matter on the one side and controllable and durable hard materials and electronics on the other side.

Using lipids for a variety of soft-matter, low-technological applications is not new. Conventionally, lipids are used as surfactants in foodstuffs and emulsions, for cosmetics, for surface coatings, and for simple formulations of drugs. Other classical applications include flotation, lubrication, and foam-formation. More modern applications use subtle chemical differences in the head-group nature of lipids for surface covering of implants to avoid blood-protein adsorption or for micro-beads in assays for protein separation.

However, these applications are far from exhausting the applicability of lipids. In particular, they do not exploit the fact that lipids were evolved by nature to function specifically on the micro- and nanometer scale, and they do not fully appreciate the fantastic and delicate structural and functional properties that soft lipid-based materials can have. Making smart and intelligent nanotechnology out of lipids is, not however, an easy task. It involves many intellectual and technical challenges, but it also holds some magnificent promises. Some lipid-based technologies may in fact turn out to be important for the future of mankind and prove helpful for survival on the planet.

In this chapter we shall describe selected examples of how the insights we have obtained throughout this book on lipids, lipid structure, and lipid functionality may be used for smart micro- and nano-encapsulation technologies, making new processing technologies, designing functional surfaces for micro- and nano-electronics, formulating drugs, and developing intelligent drug-delivery systems for handling serious diseases like cancer. The examples are chosen with the view of illustrating that surprising new technologies can emerge by using lipids for the kind of tasks they are good at.

Biological materials, in particular, membrane assemblies, are good at identifying chemical compounds and turning the identification into a signal. They can also amplify the signal by cascade processes, turn a chemical signal into a physical process and vice versa, and transport material and energy from one location to another in a very controlled fashion. All these properties are what you want from electronic components like amplifiers, switches, gates, storage units, and sensors. It is, therefore, not unexpected that some of the first and promising uses of lipid-based technologies include sensors for nano-biotechnology.

One area of biosensor nanotechnology deals with biological receptor molecules that on a proper support can perform physical sensing and amplification. For this purpose, a self-assembled lipid bilayer serves both as an extremely thin electric insulator and as an embedding medium for the incorporation of biological receptors. The bilayer, which must suppress the non-specific binding of ligands and leave the sensing for the receptors, is typically supported on a solid wafer made of metal or semiconductor materials. The lipid bilayer is separated from the hard support by an appropriate cushion, often made of long-chain molecules and polymers that are chemically linked to the support. A schematic illustration of a supported lipid membrane on a soft polymer cushion is shown in Fig. 20.1. The cushion is necessary for providing the incorporated receptor molecules with a biomimetic environment. It makes room for the hydrophilic part of the receptor that protrudes from the membrane. The cushion also provides for a compartment that can contain a substantial amount of water and ions. The self-healing property of the soft lipid bilayer is important for maintaining the bilayer as an insulator with

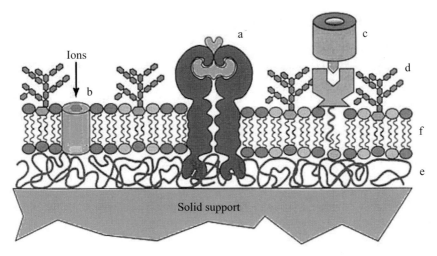

Fig. 20.1. Schematic illustration of a supported lipid membrane (**f**) on a soft polymer cushion (**e**) that separates the bilayer from a solid support or substrate. In the bilayer are incorporated (**a**) biological receptors that are trans-membrane proteins, (**b**) ion channels, (**c**) proteins that can activate enzymes, and (**d**) lipopolymers and glycolipids

as few defects as possible. The binding of a ligand to the receptor is monitored either by a capacitance or electrical current measurement across the layered device. The monitoring of the signal can be enhanced by employing an electric-optical transducer.

The Australian biophysicist Bruce Cornell has developed a new concept for a sensor technology based on supported lipid bilayers of the type in Fig. 20.1. The concept involves using an ion channel and a biophysical principle for sensing. The ion channel is gramicidin A, as we discussed in Sect. 17.4. The channel can be made conductive if two monomers of gramicidin A match up across the bilayer. The idea is now to tether gramicidin monomers in the lower monolayer leaflet of the bilayer to the solid support. Gramicidin monomers in the upper leaflet are linked to specific chemical groups that can bind to the molecules that have to be sensed. Depending on the detailed chemistries used, it is possible to open and close gramicidin channels in such a way that the resulting channel current can be employed to monitor substances in extremely small concentrations, down to one part in a thousand billions. Future sensors based on similar principles may involve other pore-forming transmembrane objects such as amphiphilic peptide antibiotics or toxins. One example is α-hemolysin, a toxin related to antrax and cholera, which is produced by the bacterium *Staphylococcus aureus*. Variants of α-hemolysin can used to form pores in supported lipid bilayers that can be triggered to open or close by external chemical and physical stimuli.

Sensors based on supported lipid bilayers can be used to monitor proteins, DNA, hormones, drugs, and other chemicals. The advantages of this type of sensor are that they can be made very small and that they can detect and quantitatively measure extremely small amounts of material. They are expected to find use for medical diagnostics and environmental monitoring, and they may well come to play a role in a futuristic computer technology based on biological processes.

Another line of technological application of functionalized supported lipid bilayers like in Fig. 20.1 uses whole cells or phantom cells that adhere to the surface. Phantom cells could be large liposomes incorporated with appropriate lipids, cytoskeletal polymer networks, and surface-cell receptors. Such supported cells can be used to monitor cell adhesion, cell motility, and locomotion, and they can serve as a laboratory for investigating growth conditions for cell cultures on surfaces.

Several proteins act as molecular motors transforming ATP into mechanical work. Some of these motors function in association with membranes and biological fibers providing for cellular motility. This intricate molecular machinery still awaits biomedical and nanotechnological applications, e.g., as nano-actuators. Some membrane-bound proteins rotate, like the integral membrane protein ATP synthase, which pumps protons while it rotates. Others, like myosin, are involved in muscle contraction by exerting force on the actin filaments. Kinesin acts as a molecular motor at microtubules and helps organize membrane compartments and facilitates vesicle trafficking in the cell. This motor plays a key role during cell division when the chromosomes have to be separated. Kinesin is also involved in the motion of the outer cell hairs and flagella of micro-organisms and is therefore, in control of the motility of these cells.

Evan Evans has demonstrated that it is possible to form nanometer-scale conduits and networks from lipids in fluid bilayers. The networks are composed of straight lipid bilayer tubes of controllable diameters in the range of 20–200 nm. Examples of such networks are shown in Fig. 20.2. These networks may form the starting point for novel lipid-based technologies, for small-scale confined reaction chambers, and for nano-electronics. Moreover, they are likely to turn out to be interesting assays for studying intercellular communication and transport, as well as chemical computations.

The basis for the formation of such nano-tubes and networks is the fluid and soft character of lipid bilayers in vesicles and liposomes. The tubes can be pulled out from the surface of vesicles by micro-manipulation techniques, as shown in Fig. 20.2a. Due to the fluidity of the bilayer and the fact that it is controlled by a surface tension, the tube that is formed is linear and is attached at the shortest distance between the mother liposome and the pulling device. The angles in a network of tubes, as shown in Fig. 20.2b, can be controlled in this way.

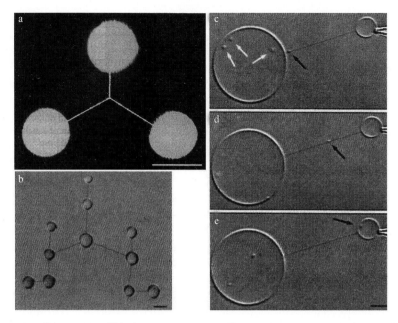

Fig. 20.2. Formation of lipid nano-tubes by pulling on liposomes in the liquid phase using micro-pipette techniques. (**a**) Fluorescent image of a vertex of three connected lipid nano-tubes. The scale bar is 10 μm. (**b**) Nano-tubular network connecting eleven liposomes. The scale bar is 10 μm. (**c**)–(**e**) Two liposomes connected by a single nano-tube. A small internal liposome of diameter 150 nm inside the large liposome is transported through the tube to the small liposome by applying a tension by the micro-pipette on the small liposome in the upper right corner. The experiment illustrates how material can be transported within a connected lipid nano-tubular network. The scale bar is 5 μm

Due to the self-assembly character of the lipid bilayers, the diameter of the nano-tubes can be controlled by varying lipid material, membrane tension, or environmental conditions in the solution. The tubular network can be used as a template for forming solid structures by using photo-induced cross-linking and polymerization of suitable macromolecular monomers in the solution. A further metallization of the network will lead to a conducting network. The advantage of this approach, in contrast to forming nano-conduits of initially hard materials like carbon nano-tubes, is that the final dimensions of the lipid-based structures and their connectivity can be widely varied and very accurately controlled. Under proper conditions, lipid tubules can also form spontaneously in solution, leading to a disordered assembly of tubes. This much less well controlled system can also be used as a template for metallization. The resulting solid tubes can be used as depots for the slow release of chemicals and have, e.g., been employed together with paints to prevent marine fouling on ships.

Liposomes connected by a nanoscopic tubular connection like in Fig. 20.2a can be considered a miniature set of connected chemical reaction containers, each containing a liquid volume that can be smaller than one part in a million billions of a liter. This is the proper realm for a study of the biochemical reactions of the cell. An example of a transport process involving the transfer of a small vesicle from one liposome to another through the narrow nano-tube is illustrated in Fig. 20.2c–e. The process is initiated by creating a surface tension difference between the two liposomes by injecting a proper solution into one of the liposomes.

20.2 Lipids Deliver Drugs

In order to reach their target, all drugs, in one way or another, have to be transported in aqueous environments and have to cross biological barriers of hydrophilic and/or hydrophobic character. The major barriers are the skin, the gastrointestinal epithelium, and the blood-brain barrier. This immediately poses a host of problems that pharmacists have to face when designing and formulating new drugs. Many potent drugs are hydrophobic or amphiphilic since their targets are membranes and membrane-bound receptors. Such drugs are not easy to administer, since they do not readily dissolve in the blood stream or the juices in the gastrointestinal tract. They have to be introduced in a formulation that improves their solubility. In fact, many very promising and potent drugs have never made it to the clinic because they are too hydrophobic to be prepared in a suitable formulation.

Other potent hydrophilic or hydrophobic drugs may be difficult to apply because they are solid or they become degraded too quickly or long before they reach their targets. Such drugs need either to be encapsulated for protection or incorporated into a formulating agent or depot that provides for sustained, retarded, or controlled release of the drug. Even if such provisions are made, other problems may remain or be induced by the formulation. The drug may not be able to cross the necessary plasma or intercellular membranes to reach its target, it may have too low bioavailability, or the pharmacokinetics may have been altered unfavorably by the formulation.

Lipids and lipid encapsulation technologies may be the solution to some of these problems. First of all, lipids are amphiphiles designed to mediate hydrophobic and hydrophilic environments, which makes them perfect emulsifiers. Secondly, many lipids are biocompatible and biodegradable and hence harmless to biological systems. Thirdly, lipids are a rich class of molecules allowing for a tremendous range of possibilities. Finally, and possibly most important, lipids are the stuff out of which the barriers that limit drug transport and delivery are themselves made. Therefore, by using lipids for transport and delivery of drugs, one can exploit nature's own tricks to interact with cells, cell membranes, and receptors for drugs. In addition to this, lipids or derivatives of lipids may themselves act as potent drugs or agents that influence cell functioning and signalling pathways (cf. Sect. 19.3).

In Chap. 17, we described how lipid membranes and lipid-bound proteins are targets for drugs. Knowledge about the molecular organization as well as the transverse and lateral structure of membranes is therefore of seminal importance for understanding how drugs bind to, penetrate, and possibly diffuse across biological membranes. Many drugs are peptides, proteins, hormones, or nucleotides, and their interactions with membranes are intimately controlled by the properties of lipids. Moreover, the mechanisms of action of many drugs involve subtle membrane-controlled triggering mechanisms, fusion events, and complex structural transitions.

An unconventional application of lipids and their polymorphism for drug-delivery purposes exploits the fact that some lipids in dispersed micellar or lamellar phase undergo a transition to a non-lamellar phase upon hydration or by increasing the temperature as described in Sect. 4.3. Whereas non-lamellar phases, such as inverted hexagonal and cubic phases, are usually of little use as drug-delivery systems because they are three-dimensional and non-particulate, they can instead be used as drug depots. As an example, local anesthetics such as lidocaine can be made to release over longer periods than after simple injection by formulating them with lipids that are in a dispersed lamellar phase at room temperature. Upon injection of the dispersed formulation into the tissue, or application to an internal wound caused by surgery, the higher body temperature takes the formulation from a liquid dispersed phase into the cubic phase, which is much more viscous. This leads to a beneficial slower and prolonged release of the drug. Another example is a controlled-release formulation of an antibiotic, metronidazole, against periodontitis. This antibiotic can be administered in a dehydrated micellar phase that swells into an inverted hexagonal phase when injected in the gingiva near the tooth. From there it is slowly released.

Some drugs are themselves fats and lipids, or derivatives thereof, e.g., lyso-ether lipids and eicosanoid derivatives like anandamide. In Chap. 16, we discussed the importance of polyunsaturated lipids in the form of EPA and DHA for human brain development and function. Disorder in the phospholipid spectrum of brain cells has been associated with the occurrence of mental and bipolar diseases such as autism, schizophrenia, and manic-depression. David Horrobin has recently advocated that schizophrenia may be treated with EPA. EPA also seems to have potential for treating some cancers by suppressing the synthesis and expression of growth-regulatory proteins, including cyclines, that are upregulated in the cancer cells.

In Sect. 20.4, we shall describe a class of ether lipids and lipid derivatives that can act as anti-cancer drugs in a special liposomal formulation where the active drug and the drug carrier are two sides of the same object.

20.3 Liposomes As Magic Bullets

One of the key problems in the treatment of serious diseases is that many potent drugs are very poisonous and not only kill the diseased cells but also healthy ones. In the beginning of the twentieth century, the father of modern medicinal chemistry, Paul Erlich, envisioned the perfect drug as a "bullet" that automatically targets and selectively kills the diseased cells without damaging healthy tissue. The term "magic bullet" refers to this perfect drug. Dr. Erlich's magic bullet has since been the Holy Grail in medicinal chemistry.

In a modern version, a magic bullet could be represented as shown in Fig. 20.3. The magic bullet contains a number of features. First of all, it contains the drug or another related compound, a so-called prodrug, that can be turned into a drug by an appropriate mechanism. The drug is attached to a carrier, which can transport the drug to the target. Finally, the carrier may contain some kind of homing device that can search for and target the site where the drug is supposed to act.

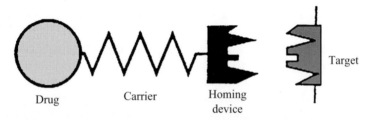

Fig. 20.3. Schematic illustration of Dr. Erlich's "magic bullet" that can target and deliver a drug to a specific site. The magic bullet consists of a drug, a carrier, and a homing device that can identify the target for the drug

Ever since the British haematologist Sir Alec Bangham in the early 1960s identified liposomes as small water-containing lipid capsules, it has been a dream to use liposomes as magic bullets for drug delivery. Liposomes appear ideal for this purpose for several reasons. They are made of biocompatible, nontoxic, and biodegradable materials; they have an aqueous lumen that can contain hydrophilic substances; they are composed of a lipid bilayer that can accommodate hydrophobic or amphiphilic drugs; they can be made in different sizes, some of them small enough to travel into the finest capillaries; and they can be associated with specific chemical groups at the liposome surface that can act as homing devices and thereby target specific cells. An illustration of a liposome magic bullet is shown in Fig. 20.4a.

Despite tremendous efforts made by a large number of researchers to devise liposome-based drug-delivery systems, it is only in recent years that some success seems to be within reach. One of the major problems has been that conventional liposomes injected into the blood stream quickly become

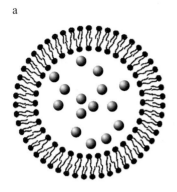 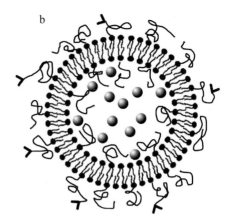

Fig. 20.4. Liposomes as magic bullets for transport and delivery of drugs. (**a**) First-generation liposome with possible hydrophilic drugs enclosed in the aqueous lumen. Hydrophobic and amphiphilic drugs can in addition be incorporated into the lipid bilayer. A typical diameter of the liposome is 100 nm. (**b**) Second-generation liposome, the so-called "stealth liposome," with a surface covered by long-chain polymeric molecules that protect the liposome from being captured by the human immune system. Some of the polymers can further be linked to a molecular group that can actively target specific cells. These targeting groups function as the homing device in the magic bullet construct in Fig. 20.3

captured and degraded by the macrophages of immune system. When that happens, the drug is released in the blood where it becomes degraded or, even worse, may damage or kill the red blood cells. Conventional liposomes, therefore, seldom make it to other sites in the body than the liver and the spleen.

A major step forward was made with the invention of the second-generation liposomes, the so-called "stealth liposomes," that are screened from the macrophages by a polymer coat as illustrated in Fig. 20.4b. Similar to the cover of a Stealth airplane, which makes the plane invisible to radar systems, the polymer coat of the stealth liposome makes it invisible to the body's immune system. This coat is constructed by incorporating a certain fraction of lipopolymers into the liposome. A lipopolymer is a lipid molecule to whose head group is chemically linked a long-chain polymer molecule that is water soluble. The aqueous polymer coat exerts several physical effects. One is to provide an entropic repulsion (cf. Sect. 5.3) between different liposomes and between liposomes and the special proteins that usually adsorb to foreign particles in the blood as part of the immune system's defense strategy. Another effect is that the water-soluble polymers make the surface of the liposome look like harmless water. The stealth liposomes exhibit a circulation time in the blood that is far longer than that of bare conventional liposomes.

The increased stability of the stealth liposomes implies that they can retain their poisonous load from the blood and have time to reach diseased sites before they eventually are cleared by the macrophages. Surprisingly and very fortunately, the stealth liposomes are found to passively target sites of trauma. The reason for this fortuitous mechanism is that the liposomes, due to their small size and their long circulation times, can venture into the leaky capillaries that are characteristic of tissues infested with tumors, inflammation, and infections. The diseased tissue sucks up the circulating liposomes in its porous structure. This enhances the efficacy of the drug and limits severe side effects.

One of the first drugs that was successfully used in a liposomal formulation was amphotericin B. Amphotericin B is a very potent antibiotic used in the treatment of systemic infections that are very serious for immunodepressed patients such as AIDS patients and patients undergoing chemotherapy. Amphotericin B is extremely toxic but water soluble and can, therefore, be readily encapsulated in the aqueous lumen of the liposome. Another example is liposomal formulations of doxorubicin, which is a potent anti-cancer drug used, e.g., in the treatment of breast cancer in women. Doxorubicin is hydrophilic, and it is incorporated in the carrier liposomes as a small solid crystalline particle. A few other liposomal formulations with anti-cancer drugs and vaccines have been approved for use in patients, and there are currently a number under development and clinical testing.

An interesting example of a liposomal drug delivery system has been developed by the Slovenian-German medical biophysicist Gregor Cevč who has designed a particular type of flexible liposomes, so-called transfersomes. Transfersomes are ultra-flexible liposomes that are soft enough to squeeze through the dermal barrier of the skin, cf. Sect. 19.1. The flexibility corresponds to bilayer bending modules in the range of the thermal energy $\kappa \succeq k_B T$. The driving force for the transfer across the skin is supposed to be the gradient in water chemical potential that changes significantly going from the dry surface to the moist dermis. The details of the mechanism by which transfersomes penetrate the dermal barrier are not known. It has been suggested that the transfersomes travel along water-filled cracks in the skin and make it to very deep layers where they eventually arrive in the blood via the lymphatic system. Hence, transfersomes may be used not only to carry drugs to the skin for treatment of skin diseases, but possibly also for systemic delivery of other drugs to the whole body without requiring injections.

Recent years have witnessed a tremendous activity in the use of liposomes for carrying DNA fragments (plasmids) to be delivered at the cell nucleus. Gene delivery and gene therapy of this type, without using viral vectors, offer a large number of opportunities. Since DNA is negatively charged, liposomes containing positively charged lipids are most often used for the encapsulation, although it has been demonstrated that DNA can also condense and intercalate in stacks of neutral lipid bilayers. The positively charged liposomes, furthermore, interact more strongly with the target cells that usually are negatively charged. Gene delivery by liposomes is still a field in its infancy. However, the prospects involve the treatment of cancer and genetic diseases like cystic fibrosis.

The properties of a liposomal carrier system can be optimized to the actual case by modulating the lipid composition of the liposomes. The lesson so far has been that it is necessary to go through an elaborate optimization procedure that takes a large number of details into account with respect to the actual drug, the actual disease, and the molecular composition of the liposomal formulation. With respect to active targeting of liposomes, only limited progress has been made so far. Hence, the full realization of Dr. Erlich's magic bullet in Fig. 20.3 by means of liposomes remains a visionary idea.

Paradoxically, one of the outstanding problems is not so much how to stabilize liposomes with encapsulated drugs as it is how to destabilize them and arrange for the liposomes to release and deliver a sufficiently large part of their load exactly where it is needed. An additional requirement is that the release should take place over a time span that is tuned to the mode of action of the drug. Below, in Sect. 20.4, we shall describe a couple of cases where the insight into the physics and physical chemistry of lipid bilayers and liposomes, in particular with respect to thermal phase transitions and enzymatic degradation of lipids, has provided a key to solving the problem of site-specific drug release.

20.4 Lipids Fighting Cancer

One of the problems using liposomes for cancer therapy is that, although it is possible to encapsulate the drug, e.g., doxorubicin, and thereby significantly reduce the toxic side effects of the chemotherapy, the liposomes do not necessarily deliver more drug to the tumors than through the application the free drug. The reason is that the drug cannot get out of the capsule sufficiently rapidly and in sufficiently large local doses.

A team of scientists and medical doctors at Duke University may have solved this problem using hyperthermia, that is, heating the tumor a few degrees above body temperature. It turns out that heating has several beneficial effects. The heat opens the tiny blood vessels in the tumor, making it possible for the liposomes to sneak in. Moreover, the heat enhances the uptake of the drug into the cancer cells and increases the damage that the drug does to the DNA of the cancer cell. The crucial point, however, is that the liposomes used by the Duke researchers are poised to become leaky, at temperatures a few degrees above the body temperature. The mechanism to do so is a lipid phase transition of the type we discussed in Sect. 9.2. At the transition, the lipid bilayer becomes leaky, as shown in Fig. 12.1a, and the encapsulated material flows out.

The American material scientist David Needham has used this phase transition phenomenon to construct liposomes whose drug-release mechanism is precisely the lipid phase transition. By composing the liposomes from lipids that have a phase transition and become leaky slightly above body temperature but otherwise are fairly tight at lower temperatures, Needham has succeeded in making a formulation for chemotherapy that can deliver as much as thirty times more drug at the tumor site than a conventional liposome. Figure 20.5 shows how sensitive the release in this system can be tuned to temperatures above body temperature in a range which can be clinically achieved by local heating using microwave, ultrasound, or radio-frequency radiation. The release of the drug from the heated liposomes is very fast, within twenty seconds after heating, which is a crucial factor for the therapeutic effect. The encapsulated drug escapes extremely rapidly from the liposomes, typically a million times faster than from ordinary liposomes.

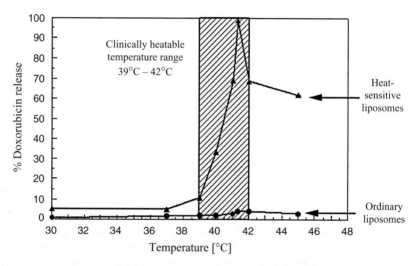

Fig. 20.5. Release of the anti-cancer drug doxorubicin from liposomes triggered by a lipid phase transition in the liposome. The liposomes sensitive to hyperthermia (*top curve*) are seen to release a very large part of the encapsulated drug over a narrow range of temperatures slightly above body temperature. In contrast, conventional liposomes (*bottom curve*) stay rather tight over the same temperature range

There is an additional benefit of using soft matter like lipid aggregates for drug encapsulation and delivery of this type. When a leaking liposome leaves the tumor area that is heated, it seals again when the temperature drops, because the liposome is a self-assembled object. The remaining drug is retained in the liposome and, therefore, does not get out into possibly healthy tissue. If this liposome later diffuses back into the heated tumor area, more drug can get released. The system developed at Duke University has shown some very promising results in the treatment of breast cancer and may eventually also be used for other cancers.

In order to use hyperthermia as a mechanism for drug release in chemotherapy it is necessary to know which area to heat. The position and the size of the diseased tissue, therefore, have to be known beforehand. These conditions may not be fulfilled for many cancers, particularly in their early stages of development. In order to come closer to Dr. Erlich's vision of a magic bullet, it would be desirable to have a liposome that itself could identify the sites of disease and by some appropriate automatic mechanism be triggered to unload the drug precisely at those sites.

For this purpose, it may be possible to use specific phospholipases to automatically trigger the opening of liposomes at diseased sites. It is known that certain variants of secretory phospholipase A_2 are over-expressed in malignant tumors and sometimes occur in a concentration that is maybe ten times larger than in healthy tissue. As we saw in Sects. 12.3 and 18.1, phospholipase A_2 catalyzes the hydrolysis of phospholipids into lysolipids and free fatty acids, leading to a leakage of bilayers and eventually to a breakdown of liposomes. Moreover, the activity of these enzymes is tightly regulated by the physical properties of the lipid bilayer. Hence, by tailoring liposomes to be sensitive to enzymatic breakdown under circumstances prevailing in the tumor, on the one side, and by taking advantage of the elevated levels of phospholipase A_2 in the tumor, on the other side, a smart principle of automatically triggered drug release suggests itself.

These examples show that lipids in the form of liposomes may be of great help in fighting cancer. But the role of lipids does not stop with that. In 1999, the Danish pharmacist Kent Jørgensen realized that it should be possible to use the phospholipase-induced drug release by the mechanism described above not only to release drugs, but also to produce a potent drug at the site of disease. The idea is amazingly simple and illustrated schematically in Fig. 20.6. The trick is to use liposomes made of lipids that upon hydrolysis via the phospholipase lead to products that themselves are drugs. Compounds that can be turned into drugs but are not drugs themselves are called prodrugs. The prodrug in this case is a lipid in which the fatty-acid chain in the first position is bound to the glycerol backbone by an ether bond and in the second position by an ester bond. After hydrolysis catalyzed by phospholipase A_2, cf. Fig. 18.1, the products are a lyso-ether lipid and a free fatty acid. The lyso-ether lipid is an extremely potent anti-cancer drug that so far has found limited use in conventional chemotherapy because it kills red blood cells. However, in its masked prodrug form as part of a lipid, it turns out to be completely harmless. Hence, the prodrug can be incorporated into long-circulating stealth liposomes that, upon accumulation in the capillaries of porous cancerous tissue, are broken down by phospholipase A_2. The drug is therefore produced exactly where it is needed, in fact, without any

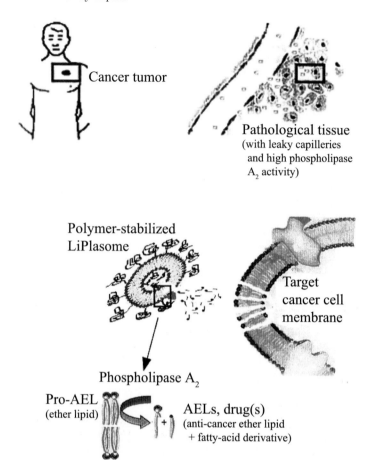

Fig. 20.6. Schematic illustration of lipids fighting cancer. LiPlasomes, which are liposomes with ether lipids, circulate in the blood stream for a long time because they have a polymer coat. The LiPlasomes accumulate in cancerous tissue by a passive mechanism caused by the leaky capilleries characteristic of solid tumors. The LiPlasomes are broken down in the tumor by the enzyme phospholipase A_2, which is upregulated in cancer. The products, anti-cancer lyso-ether lipids (AEL) and free fatty acid, are released and transported into the cancer cell

prior knowledge of the localization of the tumor. The drug carrier and the drug are in this system two sides of the same thing. In fact, the lipid drugs self-assemble into their own carrier liposome. Obviously, liposomes of this type, which are called *LiPlasomes*, can also be made to include conventional anti-cancer agents like doxorubicin and cisplatin. This may prove useful in combination therapies. One could also imagine that the second hydrolysis product, the fatty acid, is chosen to have some additional therapeutic effect.

But this is not the full story of ether lipids fighting cancer. Firstly, cancer cells are very vulnerable to ether lipids because they do not contain enzymes that can break down ether lipids. Except for red blood cells, other healthy cells do have such enzymes. Secondly, there are a number of added benefits from the LiPlasome concept. As an example, the lysolipids and the free acids produced by the phospholipase-triggered hydrolysis near the cancer cells have some beneficial effects via their propensity for forming non-lamellar lipid phases, as described in Chap. 4. They facilitate the transport of the drug into the cancer cell by lowering the permeability barrier of the target cell. Compounds with this capacity are called drug enhancers. In the case of lysolipids and fatty acids, the mode of action of the enhancers is based on a purely physical mechanism caused by the effective conical shapes of the molecules. In this way, the LiPlasome not only carries a prodrug, it also carries proenhancers that are turned into enhancers at the target.

Epilogue: Fat for Future

In her beautiful 1998 book *The Fats of Life*, Caroline Pond writes, "Genes contain the information to build the cell, proteins catalyze the necessary chemical reactions, but phospholipids act as the marshals, holding the biochemical machinery together and helping to maintain the right chemical environment." To this we could add that the marshalling task of the lipids also involves intricate signalling and an amazing ability to support and carry out function. Although the blueprint for the production of all the essential molecules of the cell is provided by the genes, the actual building of the cell and the assembly and functioning of all its molecular machinery are not written in the genes. These phenomena are based on self-organization processes controlled by the laws of physics. In these processes, the lipids play a key role that often has been overlooked or forgotten.

The main aim of this book has been to demonstrate how nature chose a wonderful and versatile class of molecules, the lipids, to structure and organize living matter in a way that provides for unique functions. Lipids and fats are not only foodstuff. Life as we know it could not have evolved or been sustained without lipids. Obviously, revealing the design principles underlying the functioning of lipids in living systems not only provides fundamental insights into biology and the evolution of life, it also holds a strong promise for translating the obtained insights into technologies useful for improving our life conditions.

Lipid-based technologies are inspired by nature's own nanotechnologies that have been developed over evolutionary time scales to optimize biological function on the cellular and subcellular levels, i.e., precisely on the nanometer scale. The design and function of nature's materials are a full-blown nano-science using "bottom-up" design principles leading to soft materials of unique function and durability. The study of natural materials and their function is a truly multidisciplinary endeavor. The nano-science based on lipids and other biological molecules operates in a domain where the boundaries between traditional disciplines of science – physics, chemistry, and biology – no longer makes sense. It is in this domain where we may in the future expect real innovative developments within drug discovery and design.

Technology based on lipids and self-assembly processes can also possibly help meet the increasing needs for future sustainability of industrial processes.

As natural materials, lipids are biodegradable and can be reused and recycled. Use of nature's own bottom-up principles for tomorrow's nanotechnology and nano-electronics basically puts the factory in a beaker. It does not necessarily presuppose a billion-dollar factory with expensive clean-room facilities that is required for conventional micro- and nano-electronics based on semiconductor chip technology.

There are some issues that make studies of fat and lipids an urgent matter for mankind. We are currently witnessing a rapid increase in non-communicable diseases such as obesity, type II diabetes, hypertension, cardio-vascular diseases and stroke, colon and breast cancers, mental ill-health, as well as perinatal conditions. Obesity is increasing globally almost like an epidemic. The burden of ill-health and the number of deaths due to these diseases are now greater than for all infectious diseases combined. The rise in mental ill-health follows the rise in cardio-vascular diseases. It is particularly troublesome that this rise in mental ill-health and behavioral problems is largest in young people. It indicates that a well-protected system like the brain is being affected.

In a paradoxical way, the human genome project actually indirectly showed that the rise in these non-communicable diseases is not genetically determined. Our genes have not changed over the few decades we are talking about. The major cause has to be changes in the diet and environmental conditions. So the question is then how the diet and, in particular, the dietary fats, are involved in the regulation and expression of genetic information. This is again where lipids and biological membranes get involved.

It has been known for some time that the physical non-communicable diseases belong to the so-called metabolic syndrome and are therefore linked to our diet. Special diets like the traditional fish-rich Icelandic and Japanese diets and the special olive-oil rich Mediterranean diet are well-known for promoting longevity and low incidence of heart attacks. It is now gradually becoming clear that a number of mental diseases such as schizophrenia, manic-depression, Alzheimer's, Parkinson's, and autism may also be related to changes in the diet. The polyunsaturated fats and their derivatives such as the cell-regulating eicosanoids become key issues here. The general failure of governmental programs aimed at fighting diseases related to the metabolic syndrome, such as coronary heart diseases and obesity, by focusing on fats and basically neglecting carbohydrates in the diet, has highlighted the need for a balanced view of which roles fats and lipids actually play for life.

A large number of mysteries concerning the role of fats and lipids for life are still unresolved. Some of the more obvious mysteries involve the role of lipids in evolution, the need for lipid diversity in membrane function, the physical principles that control cell signalling by lipids, and the relationship between nutrition and health.

Many more mysteries are likely to turn up as we realize how life is a matter of fat.

Bibliography

The Bibliography provides references to books, review papers, and research articles accounting for most of the factual statements made in this book. It begins with a list of general books and a collection of review papers and is followed by separate bibliographies for each chapter of the book.

GENERAL BOOKS AND REVIEWS ON MEMBRANES AND LIPIDS

Bloom, M., E. Evans, and O. G. Mouritsen. Physical properties of the fluid-bilayer component of cell membranes: a perspective. Q. Rev. Biophys. **24**, 293–397 (1991).

Boal, D. *Mechanics of the Cell.* Cambridge University Press, Cambridge (2002).

Cevc, G. (ed.). *Phospholipids Handbook.* Marcel Dekker, Inc., New York (1993).

Cevc, G. and D. Marsh. *Phospholipid Bilayers. Physical Principles and Models.* John Wiley and Sons, New York (1987).

Cotterill, R. M. J. *Biophysics. An Introduction.* John Wiley and Sons, New York (2002).

Deamer, D. W., A. Kleinzeller, and D. M. Fambrough (eds.). *Membrane Permeability. 100 Years Since Ernest Overton.* Academic Press, London (1999).

Disalvo, A. and S. A. Simon (eds.). *Permeability and Stability of Lipid Bilayers.* CRC Press, Inc., Boca Raton, Florida (1995).

Epand, R. (ed.). *The Properties and Biological Roles of Non-lamellar Forming Lipids.* Chem. Phys. Lipids. (Special Issue) **81**, 101–264 (1996).

Finegold, L. (ed.). *Cholesterol in Membrane Models.* CRC Press, Inc., Boca Raton, Florida (1993).

Gennis, R. B. *Biomembranes. Molecular Structure and Function.* Springer Verlag, Berlin (1989).

Gurr, M. L. *Lipids in Nutrition and Health: A Reappraisal.* The Oily Press, Bridgwater (1999).

Hanahan, D. J. *A Guide to Phospholipid Chemistry.* Oxford University Press, New York (1997).

Jackson, M. B. (ed.). *Thermodynamics of Membrane Receptors and Channels.* CRC Press, Inc., Boca Raton, Florida (1993).

Katsaras, J. and T. Gutberlet (eds.). *Lipid Bilayers. Structure and Interactions.* Springer-Verlag, Berlin (2001).

Kinnunen, P. K. J. (ed.). *Peripheral Interactions on Lipid Surfaces: Towards a New Biomembrane Model.* Chem. Phys. Lipids (Special Issue) **101**, 1–137 (1999).

Kinnunen, P. K. J. and O. G. Mouritsen (eds.). *Functional Dynamics of Lipids in Biomembranes.* Chem. Phys. Lipids (Special Issue) **73**, 1–236 (1994).

Larsson, K. *Lipids. Molecular Organization, Physical Functions, and Technical Applications.* The Oily Press, Dundee (1994).

Lasic, D. and Y. Barenholz (eds.). *Nonmedical Applications of Liposomes.* CRC Press, Inc., Boca Raton, Florida (1995).

Lee, A. G. and P. L. Yeagle (eds.). *Membrane Protein Structure.* Biochim. Biophys. Acta (Special Issue) **1565**, 144–367 (2002).

Léger, C. L. and G. Béréziat. *Biomembranes and Nutrition.* Inserm, Paris (1989).

Lipowsky, R. and E. Sackmann (eds.) *Handbook of Biological Physics* (A. J. Hoff, series ed.) *Vols. IA and IB: Structure and Dynamics of Membranes, from Cells to Vesicles.* Elsevier Science B. V., Amsterdam (1995).

Luisi, P. L. and P. Walde (eds.). *Giant Vesicles.* John Wiley and Sons Ltd., New York (2000).

Mertz Jr., K. M. and B. Roux (eds.). *Membrane Structure and Dynamics. A Molecular Perspective from Computation and Experiment.* Birkhäuser Publ. Co., New York (1996).

Mouritsen, O. G. and O. S. Andersen (eds.). *In Search of a New Biomembrane Model.* Biol. Skr. Dan. Vid. Selsk. **49**, 1–224 (1998).

Petrov, A. G. *The Lyotropic State of Matter: Molecular Physics and Living Matter Physics.* Gordon and Breach, Amsterdam (1999).

Pond, C. M. *The Fats of Life.* Cambridge University Press, Cambridge (1998).

Preston, T. M., C. A. King, and J. S. Hyams (eds.). *The Cytoskeleton and Cell Mobility.* Blackie and Son Ltd, London (1990).

Robertson, R. N. *The Lively Membranes.* Cambridge University Press, Cambridge (1983).

Rosoff, M. (ed.). *Vesicles.* Surfactant Science Series, Vol. 62. Marcel Dekker, Inc., New York (1996).

Safran, S. A. *Statistical Thermodynamics of Surfaces, Interfaces, and Membranes.* Addison-Wesley Publ. Co., New York (1994).

Schnitsky, M. (ed.). *Biomembranes. Physical Aspects.* Balaban Publ., VCH, New York (1993).

Silver, B. L. *The Physical Chemistry of Membranes.* Allen and Unwin, Winchester (1985).

Sleytr, U. B., P. Messner, D. Pum, and M. Sára (eds.). *Crystalline Bacterial Cell Surface Proteins.* Academic Press, San Diego, California (1996).

Tien, H. T. and A. Ottova-Leitmannova. *Membrane Biophysics. As Viewed from Experimental Bilayer Lipid Membranes.* Elsevier, Amsterdam (2000).

Vance, D. E. and J. Vance (eds.). *Biochemistry of Lipids, Lipoproteins, and Membranes.* Elsevier, Amsterdam (1991).

Warren, R. C. *Physics and the Architecture of Cell Membranes.* Adam Hilger, IOP Publishing Ltd, Bristol (1987).

Yeagle, P. L. (ed.). *The Structure of Biological Membranes.* 2nd Edition, CRC Press, Boca Raton, Florida (2005).

PROLOGUE:
LIPIDOMICS – A SCIENCE BEYOND STAMP COLLECTION

Lagarde, M., A. Géloën, and M. Record, D. Vance, and F. Spener. Lipidimics is emerging. Biochim. Biophys. Acta **1634**, 61 (2003).

Rilfors, L. and G. Lindblom. Regulation of lipid composition in biological membranes – biophysical studies of lipids and lipid synthesizing enzymes. Coll. Surf. B: Biointerfaces **26**, 112–124 (2002).

Spener, F., M. Lagarde, A. Géloën, and M. Record. What is lipidomics? Eur. J. Lipid Sci. Technol. **105**, 481–482 (2003).

1 LIFE FROM MOLECULES

Alberts, B., A. Johnson, J. Lewis, M. Raff, K. Roberts, and P. Walter. *Molecular Biology of the Cell.* Garland Publishing, Inc. New York (2002).

Bak, P. *How Nature Works. The Science of Self-organized Criticality.* Springer Verlag, New York (1996).

Bloom, M., E. Evans, and O. G. Mouritsen. Physical Properties of the Fluid-Bilayer Component of Cell Membranes: A Perspective. Q. Rev. Biophys. **24**, 293–397 (1991).

Branden, C. and J. Tooze. *Introduction to Protein Structure.* 2nd edition. Garland Publishing, Inc. New York (1999).

Brown, T. A. (ed.). *Genomes.* 2nd edition. John Wiley and Sons, New York (2002).

Cooper, G. M. *The Cell. A Molecular Approach.* ASM Press, Washington D.C. (1997).

Deamer, D. W. Role of amphiphilic compounds in the evolution of membrane structure on early earth. Origins Life **17**, 3–25 (1986).

Flyvbjerg, H., J. Hertz, M. H. Jensen, O. G. Mouritsen, and K. Sneppen (eds.). *Physics of Biological Systems. From Molecules to Species.* Springer-Verlag, Berlin (1997).

Gennis, R. B. *Biomembranes. Molecular Structure and Function.* Springer-Verlag, Berlin (1989).

Goodwin, B. *How the Leopard Changed Its Spots.* Phoenix Giants, London (1994).

Hilgeman, D. W. Getting ready for the decade of the lipids. Annu. Rev. Physiol. **65**, 697–700 (2003).

Howland, J. L. *The Surprising Archae.* Oxford University Press, Oxford (2000).

International Human Genome Sequencing Consortium. Initial sequencing and analysis of the human genome. Nature **409**, 860–921 (2001).

Kauffmann, S. A. *The Origins of Order. Self-organization and Selection in Evolution.* Oxford University Press, Oxford (1993).

Kinnunen, P. K. J. On the principles of functional ordering in biological membranes. Chem. Phys. Lipids **57**, 375–399 (1991).

Margulis, L. *Origin of Eukaryotic Cells.* Yale University Press, New Haven (1970).

Morowitz, H. J. *Beginnings of Cellular Life.* Yale University Press, New Haven (1992).

Mouse Genome Sequencing Consortium. Initial sequencing and comparative analysis of the mouse genome. Nature **420**, 520–562 (2002).

Nelson, P. *Biological Physics. Energy, Information, Life.* Freeman and Co., New York (2004).

Streyer, L. *Biochemistry.* 4th edition. Freeman and Co, New York (1995).

Venter, J. C. et al. The sequence of the human genome. Science **291**, 1304–1351 (2001).

Woese, C. R. Bacterial evolution. Microbiol. Rev. **51**, 221–271 (1987).

2 HEAD AND TAIL

Cevč, G. (ed.). *Phospholipids Handbook.* Marcel Dekker, New York (1993).

Gennis, R. B. *Biomembranes. Molecular Structure and Function.* Springer Verlag, Berlin (1989).

Gurr, M. L. *Lipids in Nutrition and Health: A Reappraisal.* The Oily Press, Bridgwater (1999).

Pond, C. M. *The Fats of Life.* Cambridge University Press, Cambridge (1998).

Sackmann, E. Biological membranes. Architecture and function. In *Handbook of Biological Physics* (A. J. Hoff, series ed.). *Vol. I: Structure and Dynamics of Membranes* (R. Lipowsky and E. Sackmann, eds.). Elsevier Science B. V., Amsterdam (1995) pp. 1–63.

Streyer, L. *Biochemistry.* 4th edition. Freeman and Co., New York (1995).

3 OIL AND WATER

Adamson, A. W. *Physical Chemistry of Surfaces.* John Wiley and Sons, New York (1990).

Cevč, G. and D. Marsh. *Phospholipid Bilayers. Physical Principles and Models.* John Wiley and Sons, New York (1987).

Ball, P. H_2O – *A Biography of Water.* Weidenfeld and Nicolson, The Orion Publishing Group Ltd., Orion House, London, UK (1999).

Cates, M. E. and S. A. Safran (eds.). Theory of self-assembly. Curr. Opin. Colloid. Interface Sci. **2**, 359–387 (1997).

Chen, S.-H. and R. Rajagopalan. *Micellar Solutions and Microemulsions.* Springer-Verlag, New York (1990).

Evans, E. and F. Ludwig. Dynamic strengths of molecular anchoring and material cohesion in fluid biomembranes. J. Phys: Condens. Matter **12**, A315–320 (2000).

Evans, D. F. and H. Wennerström. *The Colloidal Domain. Where Physics, Chemistry, Biology, and Technology Meet.* 2nd edition. VCH Publishers, Inc., New York (1999).

Gelbart, W. M., A. Ben-Shaul, and D. Roux (eds.). *Micelles, Membranes, Microemulsions, and Monolayers.* Springer-Verlag, Berlin (1994).

Israelachvili, I. *Intermolecular and Surface Forces.* 2nd edition. Academic Press, London (1992).

Jensen, M. Ø., O. G. Mouritsen, and G. H. Peters. The hydrophobic effect: molecular dynamics simulations of the structure and dynamics of water confined between extended hydrophobic and hydrophilic surfaces. J. Chem. Phys. **120**, 9729–9744 (2004).

Jönsson, B., B. Lindman, K. Holmberg, and B. Kronberg. *Surfactants and Polymers in Aqueous Solution.* John Wiley and Sons, New York (1998).

Leckband, D. and J. Israelachvili. Intermolecular forces in biology. Quart. Rev. Biophys. **34**, 105–267 (2001).

Merkel, R., P. Nassoy, and E. Evans. Using dynamic force spectroscopy to explore energy landscapes of receptor-ligand bonds. Nature **397**, 50–53 (1999).

Safran, S. A. *Statistical Thermodynamics of Surfaces, Interfaces, and Membranes.* Addison-Wesley Publ. Co., New York (1994).

Tanford, C. *The Hydrophobic Effect: Formation of Micelles and Biological Membranes*. Wiley and Sons, New York (1980).

Vilstrup, P. (ed.). *Microencapsulation of Food Ingredients*. Leatherhead Publ., Surrey, UK (2001).

4 LIPIDS SPEAK THE LANGUAGE OF CURVATURE

Epand, R. (ed.). *The Properties and Biological Roles of Non-lamellar Forming Lipids*. Chem. Phys. Lipids. (Special Issue) **81**, 101–264 (1996).

Cullis, P. and B. de Kruijff. Lipid polymorphism and the functional roles of lipids in biological membranes. Biochim. Biophys. Acta **559**, 388–420 (1979).

Gruner, S. M. Intrinsic curvature hypothesis for biomembrane lipid composition: a role for non-bilayer lipids. Proc. Natl. Acad. Sci. USA **82**, 3665–3669 (1985).

Hyde, S., S. Andersson, K. Larsson, Z. Blum, T. Landh, S. Lindin, and B. W. Ninham. *The Language of Shape. The Role of Curvature in Condensed Matter: Physics, Chemistry, and Biology*. Elsevier, Amsterdam (1997).

Israelachvili, I. *Intermolecular and Surface Forces*. 2nd edition. Academic Press, London (1992).

Kinnunen, P. K. J. On the mechanisms of the lamellar $\to$ hexagonal H_{II} phase transition and the biological significance of H_{II} propensity. In *Nonmedical Applications of Liposomes* (D. Lasic and Barenholz, eds.). CRC Press, Inc., Boca Raton, Florida (1995) pp. 153–171.

Rilfors, L. and G. Lindblom. Regulation of lipid composition in biological membranes – biophysical studies of lipids and lipid synthesizing enzymes. Coll. Surf. B: Biointerfaces **26**, 112–124 (2002).

Österberg, F., L. Rilfors, Å. Wieslander, G. Lindblom, and S. M. Gruner. Lipid extracts from membranes of *Acoleplasma laidlawii* grown with different fatty acids have nearly constant spontaneous curvature. Biochim. Biophys. Acta **1257**, 18–24 (1995).

5 A MATTER OF SOFTNESS

Bloom, M., E. Evans, and O. G. Mouritsen. Physical properties of the fluid-bilayer component of cell membranes: a perspective. Q. Rev. Biophys. **24**, 293–397 (1991).

Boal, D. *Mechanics of the Cell*. Cambridge University Press, Cambridge (2002).

Cates, M. E., and M. R. Evans (eds.). *Soft and Fragile Matter. Non-equilibrium Dynamics, Metastability, and Flow*. Institute of Physics Publ., London (2000).

Chaikin, P. M. and T. C. Lubensky. *Principles of Condensed Matter Physics*. Cambridge University Press, Cembridge (1995).

Daoud, M. and C. E. Williams (eds.). *Soft Matter Physics*. Springer-Verlag, Berlin (1995).

de Gennes, P. G. and J. Prost. *The Physics of Liquid Crystals*. Oxford Universisty Press, Oxford (1995).

Jones, R. A. L. *Soft Condensed Matter*. Oxford University Press, New York (2002).

Petrov, A. G. *The Lyotropic State of Matter: Molecular Physics and Living Matter Physics*. Gordon and Breach, Amsterdam (1999).

Safran, S. A. *Statistical Thermodynamics of Surfaces, Interfaces, and Membranes.* Addison-Wesley Publ. Co., New York (1994).

Sackmann, E. Physics of vesicles. In *Handbook of Biological Physics* (A. J. Hoff, series ed.). *Vol. I: Structure and Dynamics of Membranes* (R. Lipowsky and E. Sackmann, eds.). Elsevier Science B. V., Amsterdam (1995) pp. 213–304.

Witten, T. A. (with P. A. Pincus). *Structured Fluids: Polymers, Colloids, Surfactants.* Oxford University Press, Oxford (2004).

6 SOFT SHELLS SHAPE UP

Boal, D. *Mechanics of the Cell.* Cambridge University Press, Cambridge (2002).

Döbereiner, H.-G. Fluctuating vesicle shapes. In *Giant Vesicles.* (P. L. Luisi and P. Walde, eds.). John Wiley and Sons Ltd., New York (2000).

Evans, E. and R. Skalak. Mechanics and thermodynamics of biomembranes. CRC Crit. Rev. Bioeng. **3**, 181–330 (1979).

Hyde, S., S. Andersson, K. Larsson, Z. Blum, T. Landh, S. Lindin, and B. W. Ninham. *The Language of Shape. The Role of Curvature in Condensed Matter: Physics, Chemistry, and Biology.* Elsevier, Amsterdam (1997).

Janmey, P. Cell membranes and the cytoskeleton. In *Handbook of Biological Physics* (A. J. Hoff, series ed.). *Vol. I: Structure and Dynamics of Membranes* (R. Lipowsky and E. Sackmann, eds.). Elsevier Science B. V., Amsterdam (1995) pp. 805–849.

Lim, G. H. W., M. Wortis, and R. Mukhopadhyay. Stomatocyte-discocyte-echinocyte sequence of the human red blood cell: evidence for the bilayer-couple hypothesis from membrane mechanics. Proc. Natl. Acad. Sci. **99**, 16766–16769 (2002).

Miao, L., U. Seifert, M. Wortis, and H.-G. Döbereiner. Budding transitions of fluid-bilayer vesicles: the effect of area-difference elasticity. Phys. Rev. E **49**, 5389–5407 (1994).

Safran, S. A. *Statistical Thermodynamics of Surfaces, Interfaces, and Membranes.* Addison-Wesley Publ. Co., New York (1994).

Seifert, U. and R. Lipowsky. Morphology of vesicles. In *Handbook of Biological Physics* (A. J. Hoff, series ed.) *Vol. I: Structure and Dynamics of Membranes* (R. Lipowsky and E. Sackmann, eds.). Elsevier Science B. V., Amsterdam (1995) pp. 403–463.

Steck, T. L. Red cell shape. In *Cell Shape: Determinants, Regulation, and Regulatory Role.* (W. Stein and F. Bronner, eds.). Academic Press, New York (1989) pp. 205–246.

7 BIOLOGICAL MEMBRANES – MODELS AND FASHION

Danielli, J. F. and H. Davson. A contribution to the theory of permeability of thin films. J. Cellular Comp. Physiol. **7**, 393–408 (1935).

Edidin, M. The state of lipid rafts: from model membranes to cells. Annu. Rev. Biophys. Biomol. Struct. **32**, 257–283 (2003).

Gennis, R. B. *Biomembranes. Molecular Structure and Function.* Springer-Verlag, New York (1989).

Gorter, E. and F. Grendel. On bimolecular layers of lipoids on chromatocytes of blood. J. Exp. Medicine **41**, 439–443 (1935).

Israelachvili, J. N. Refinements of the fluid-mosaic model of membrane structure. Biochim. Biophys. Acta **469**, 221–225 (1977).

Israelachvili, J. N. The packing of lipids and proteins in membranes. In *Light Transducing Membranes: Structure, Function, and Evolution* (D. W. Deamer, ed.). Academic Press, New York (1978) pp. 91–107.

Jacobson, K., E. D. Sheets, and R. Simson. Revisiting the fluid mosaic model of membranes. Science **268**, 1441–1442 (1995).

Mertz Jr., K. M. and B. Roux (eds.). *Membrane Structure and Dynamics. A Molecular Perspective from Computation and Experiment.* Birkhäuser Publ. Co., New York (1996).

Mouritsen, O. G. Computer simulation of lyotropic liquid crystals as models of biological membranes. In *Advances in the Computer Simulations of Liquid Crystals* (P. Pasini and C. Zannoni, eds.). Kluwer Academic Publ., Dordrecht (2000) pp. 139–187.

Mouritsen, O. G. and O. S. Andersen. Do we need a new biomembrane model? In *In Search of a New Biomembrane Model.* Biol. Skr. Dan. Vid. Selsk. **49**, 7–11 (1998).

Robertson, J. D. Granulo-fibrillar and globular substructure in unit membranes. Ann. NY Acad. Sci. **137**, 421–440 (1966).

Singer, S. and G. L. Nicolson. The fluid mosaic model of cell membranes. Science **172**, 720–730 (1972).

8 LIPIDS IN BILAYERS – A STRESSFUL LIFE

Cantor, R. S. The influence of membrane lateral pressures on simple geometric models of protein conformational equilibria. Chem. Phys. Lipids **101**, 45–56 (1999).

Cevc, G. and D. Marsh. *Phospholipid Bilayers. Physical Principles and Models.* John Wiley and Sons, New York (1987).

Fujiwara, T., K. Ritchie, H. Murakoshi, K. Jacobson, and A. Kusumi. Phospholipids undergo hop diffusion in compartmentalized cell membrane. J. Cell. Biol. **157** 1071–1081 (2002).

Katsaras, J. and T. Gutberlet (eds.). *Lipid Bilayers. Structure and Interactions.* Springer-Verlag, Berlin (2001).

Kusumi, A. and Y. Sako. Cell surface organization by the membrane skeleton. Curr. Opin. Cell Biol. **8**, 566–574 (1996).

Mertz Jr., K. M. and B. Roux (eds.). *Membrane Structure and Dynamics. A Molecular Perspective from Computation and Experiment.* Birkhäuser Publ. Co., New York (1996).

Mouritsen, O. G. and K. Jørgensen. Dynamical Order and Disorder in Lipid Bilayers. Chem. Phys. Lipids **73**, 3–26 (1994).

Nagle, J. F. and S. Tristram-Nagle. Structure of lipid bilayers. Biochim. Biophys. Acta **1469**, 159–195 (2000).

Nezil, F. A. and M. Bloom. Combined influence of cholesterol and synthetic and amphiphilic polypeptides upon bilayer thickness in model membranes. Biophy. J. **61**, 1176–1183 (1992).

Saxton, M. Lateral diffusion of lipids and proteins. In *Membrane Permeability. 100 Years Since Ernest Overton* (D. W. Deamer, A. Kleinzeller, and D. M. Fambrough, eds.). Academic Press, New York (1999) pp. 229–282.

Schmidt, T., G. J. S. Schütz, W. Baumgartner, H. J. G. Ruber, and H. S. Schindler. Imaging of single molecule diffusion. Proc. Natl. Acad. Sci. USA **93**, 2926–2929 (1996).

Sonnleitner, A., G. J. Schütz, and Th. Schmidt. Free brownian motion of individual lipid molecules in biomembranes. Biophys. J. **77**, 2638–2642 (1999).

Tieleman, D. P., S. J. Marrink, and H. J. C. Berendsen. A computer perspective of membranes: molecular dynamics studies of lipid bilayer systems. Biochim. Biophys. Acta **1331**, 235–270 (1997).

Vattulainen, I. and O. G. Mouritsen. Diffusion in membranes. In *Diffusion in Condensed Matter* (J. Kärger, P. Heitjans, and R. Haberlandt, eds.). Springer-Verlag, Berlin (2004).

White, S. H. and M. C. Wiener. The liquid-crystallographic structure of fluid lipid bilayer membranes. In *Biological Membranes. A Molecular Perspective for Computation and Experiment* (K. M. Merz and B. Roux, eds.). Birkhäuser Publ. Co., New York (1996) pp. 127–144.

9 THE MORE WE ARE TOGETHER

Domb, C. *The Critical Point*. Taylor and Francis, London (1996).

Cevc, G. and D. Marsh. *Phospholipid Bilayers. Physical Principles and Models*. John Wiley and Sons, New York (1987).

Ipsen, J. H., O. G. Mouritsen, G. Karlström, H. Wennerström, and M. J. Zuckermann. Phase equilibria in the lecithin-cholesterol system. Biochim. Biophys. Acta **905**, 162–172 (1987).

Jørgensen, K. and O. G. Mouritsen. Phase separation dynamics and lateral organization of two-component lipid membranes. Biophys. J. **95**, 942–954 (1995).

Kinnunen, P. and P. Laggner (eds.). *Phospholipid Phase Transitions*. Chem. Phys. Lipids (Special issue) **57**, 109–408 (1991).

Mouritsen, O. G. Computer simulation of cooperative phenomena in lipid membranes. In *Molecular Description of Biological Membrane Components by Computer Aided Conformational Analysis*, Vol. I (R. Brasseur, ed.). CRC Press, Boca Raton, Florida (1990) pp. 3–83.

Mouritsen, O. G. and K. Jørgensen. Dynamical order and disorder in lipid bilayers. Chem. Phys. Lipids **73**, 3–26 (1994).

de Gennes, P. G. and J. Prost. *The Physics of Liquid Crystals*. Oxford Universisty Press, Oxford (1995).

Radhakrishnan, A. and H. McConnell. Thermal dissociation of condensed complexes of cholesterol and phospholipid. J. Phys. Chem. B **106**, 4755–4762 (2002).

Riste, T. and D. Sherrington (eds.). *Phase Transitions in Soft Condensed Matter*. Kluwer Academic/Plenum Publ., New York (1990).

Vist, M. and J. H. Davis. Phase equilibria of cholesterol/dipalmitoylphosphatidylcholine. ^{2}H nuclear magnetic resonance and differential scanning calorimetry. Biochemistry **29**, 451–464 (1990).

Yeomans, J. M. *Statistical Mechanics of Phase Transitions*. Clarendon Press, Oxford (1992).

10 LIPIDS IN FLATLAND

Albrecht, O., H. Gruler, and E. Sackmann. Polymorphism of phospholipid monolayers. J. Phys. (Paris) **39**, 301–313 (1978).

Bernardino de la Serna, J., J. Perez-Gil, A. C. Simonsen, and L. A. Bagatolli. Cholesterol rules: direct observation of the coexistence of two fluid phases in native pulmonary surfactant membranes at physiological temperatures J. Biol. Chem. **279**, 40715–40722 (2004).

Bonnell, D. (ed.). *Scanning Probe Microscopy and Spectroscopy: Theory, Techniques, and Applications.* John Wiley and Sons, New York; 2nd ed. (2000).

Gaines, G. L. *Insoluble Monolayers at Liquid-Gas Interfaces.* Interscience, New York (1966).

Goerke, J. Pulmonary surfactant: functions and molecular composition. Biochem. Biophys. Acta **1408**, 79–89 (1998).

Gugliotti, M. and M. J. Politi. The role of the gel-liquid-crystalline phase transition in the lung surfactant cycle. Biochim. Biophys. Acta. **89**, 243–251 (2001).

Kaganer, V. M., H. Möhwald, and P. Dutta. Structure and phase transitions in Langmuir monolayers. Rev. Mod. Phys. **71**, 779–819 (1999).

Kim, D. H. and D. Needham. Lipid bilayers and monolayers: characterization using micropipette manipulation techniques. In *Encyclopedia of Surface and Colloid Science* (A. Hobbard, ed.). Marcel Dekker, New York (2002) pp. 3057–3086.

Knobler, C. M. Seeing phenomena in flatland: studies of monolayers by fluorescence microscopy. Science **249**, 870–874 (1990).

Larsson, K. *Lipids. Molecular Organization, Physical Functions, and Technical Applications.* The Oily Press, Dundee (1994).

McConnell, H. M. Structures and transitions in lipid monolayers at the air-water interface. Annu. Rev. Phys. Chem. **42**, 171–195 (1991).

Nalwa, H. S. *Encyclopedia of Nanoscience and Nanotechnology.* Scientific American Publ., New York (2003).

Nielsen, L. K., T. Bjørnholm, and O. G. Mouritsen. Fluctuations caught in the act. Nature **404**, 352 (2000).

Nielsen, L. K., A. Vishnyakov, K. Jørgensen, T. Bjørnholm, and O. G. Mouritsen. Nanometer-scale structure of fluid lipid membranes. J. Phys.: Condens. Matter **12**, 309–314 (2000).

Petty, M. C. *Langmuir-Blodgett Films. An Introduction.* Cambridge University Press, Cambridge (1996).

Pérez-Gil, J. Lipid-protein interactions of hydrophobic proteins SP-B and SP-C in lung surfactant assembly and dynamics. Pediatric Pathol. Mol. Med. **20**, 445–469 (2001).

Peréz-Gil, J. and K. M. W. Keough. Interfacial properties of surfactant proteins. Biochim. Biophys. Acta. **1408**, 203–217 (1998).

Piknova, B., V. Schram, and S. B. Hall. Pulmonary surfactant: phase behavior and function. Curr. Opin. Struct. Biol. **12**, 487–494 (2002).

Tamm, L. K. and Z. Shao. The application of AFM to biomembranes. In *Biomembrane Structure* (D. Chapman and P. Haris, eds.). IOS Press, Amsterdam (1998) pp. 169–185.

11 SOCIAL LIPIDS

Anderson, R. G. W. and K. Jacobson. A role for lipid shells in targeting proteins to caveolae, rafts, and other lipid domains. Science **296**, 1821–1825 (2002).

Bagatolli, L. Direct observation of lipid domains in free standing bilayers: from simple to complex lipid mixtures. Chem. Phys. Lipids **122**, 137–145 (2003).

Bergelson, L. O., K. Gawrisch, J. A. Feretti, and R. Blumenthal (eds.). Domain organization in biological membranes. Mol. Membr. Biol. **12**, 1–162 (1995).

Brown, D. A. and E. London. Structure and origin of ordered lipid domains in membranes. J. Membr. Biol. **164**, 103–114 (1998).

Coelho, F, W. L. C. Vaz, and E. Melo. Phase topology and percolation in two-component lipid bilayers: a Monte-Carlo approach. Biophys. J. **72**, 1501–1511 (1997).

Dietrich, C., L. A. Bagatolli, Z. Volovyk, N. L. Thompson, M. Levi, K. Jacobson, and E. Gratton. Lipid rafts reconstituted in model membranes. Biophys. J. **80**, 1417–1428 (2001).

Dietrich, C., B. Yang, T. Fujiwara, A. Kusumi, and K. Jacobson. The relationship of lipid rafts to transient confinement zones detected by single particle tracking. Biophys. J. **82**, 274–284 (2002).

Dufrêne Y. F. and G. U. Lee. Advances in the characterization of supported lipid films with the atomic force microscope. Biochim. Biophys. Acta **1509**, 14–41 (2000).

Edidin, M. Lipid microdomains in cell surface membranes. Curr. Opin. Struct. Biol. **7**, 528–532 (1997).

Heimburg, T. Coupling of chain melting and bilayer structure: domains, rafts, elasticity and fusion. In *Planar Lipid Bilayers and Their Applications* (H. T. Tien and A. Ottova, eds.). Elsevier, Amsterdam (2002) pp. 269–293.

Heerklotz, H. Triton promotes domain formation in lipid raft mixtures. Biophys. J. **83**, 2693–2701 (2002).

Jacobson, K. and C. Dietrich. Looking at lipid rafts? Trends Cell Biol. **9**, 87–91 (1999).

Kaasgaard, T., O. G. Mouritsen, and K. Jørgensen. Lipid domain formation and ligand-receptor distribution in lipid bilayer membranes investigated by atomic force microscopy. FEBS Lett. **515**, 29–34 (2002).

Korlach, J., P. Schwille, W. W. Webb and G. W. Feigenson. Characterization of lipid bilayer phases by confocal microscopy and fluorescence correlation spectroscopy. Proc. Natl. Acad. Sci. USA **96**, 8461–8466 (1999).

Lehtonen, J. Y. A. and P. K. J. Kinnunen. Evidence for phospholipid microdomain formation in liquid crystalline liposomes reconstituted with *Escherichia coli* lactose permease. Biophys. J. **72**, 1247–1257 (1997).

Leidy, C., W. F. Wolkers, O. G. Mouritsen, K. Jørgensen, and J. H. Crowe. Lateral organization and domain formation in a two-component lipid membrane system. Biophys. J. **80**, 1819–1828 (2001).

Loura, L. M. S., A. Fedorov, M. Prieto. Fluid-fluid membrane microheterogeneity. A fluorescence energy transfer study. Biophys. J. **80**, 776–788 (2002).

Milhiet, P. E., M.-C. Giocondi, and C. Le Grimmelec. Cholesterol is not crucial for the existence of microdomains in kidney brush-border membrane models. J. Biol. Chem. **277**, 875–878 (2002).

Mouritsen, O. G. and R. L. Biltonen. Protein-lipid interactions and membrane heterogeneity. In *New Comprehensive Biochemistry. Protein-Lipid Interactions* (A. Watts, ed.). Elsevier Scientific Press, Amsterdam (1993) pp. 1–39.

Mouritsen, O. G. and P. J. K. Kinnunen. Role of lipid organization and dynamics for membrane functionality. In *Membrane Structure and Dynamics. A Molecular Perspective from Computation and Experiment* (K. M. Mertz Jr. and B. Roux, eds.). Birkhäuser Publ. Co. (1996) pp. 463–502.

Mouritsen, O. G. and K. Jørgensen. Small-scale lipid-membrane structure: simulation vs experiment. Curr. Opin. Struct. Biol. **7**, 518–527 (1997).

Pedersen, S., K. Jørgensen, T. R. Bækmark, and O. G. Mouritsen. Indirect evidence for lipid-domain formation in the transition region of phospholipid bilayers by two-probe fluorescence energy transfer. Biophys. J. **71**, 554–560 (1996).

Pyenta, P. S., D. Holowka, and B. Baird. Cross-correlation analysis of inner-leaflet-anchored green fluorescent protein co-redistributed with IgE receptors and outer leaflet lipid raft components. Biophys. J. **80**, 2120–2132 (2001).

Ramstedt, B. and J. P. Slotte. Membrane properties of sphingomyelins. FEBS Lett. **531**, 33–37 (2002).

Rinia, H. A. and B. de Kruijff. Imaging domains in model membranes with atomic force microscopy. FEBS Lett. **504**, 194–199 (2001).

Rinia, H. A., M. M. E. Snel, J. P. J. M. van der Eerden, and B. de Kruijff. Visualizing detergent resistant domains in model membranes with atomic force microscopy. FEBS Lett. **501**, 92–96 (2001).

Silvius, J. R. Lipid modifications of intracellular signal-transducing proteins. J. Liposome Res. **9**, 1–19 (1999).

Silvius, J. R. Role of cholesterol in lipid raft formation: lessons from lipid model systems. Biochim Biophys Acta. **1610**, 174–83 (2003).

Simons, K. and E. Ikonen. Functional rafts in cell membranes. Nature **387**, 569–572 (1997).

Welby, M., Y. Poquet, and J.-F. Tocanne. The spatial distribution of phospholipids and glycolipids in the membrane of the bacterium *Micrococcus luteus* varies during the cell cycle. FEBS Lett. **384**, 107–111 (1996).

Yang, L. and M. Glaser. Formation of membrane domains during the activation of protein kinase C. Biochemistry **35**, 13966–13974 (1996).

12 LIVELY LIPIDS PROVIDE FOR FUNCTION

Disalvo, A. and S. A. Simon (eds.). *Permeability and Stability of Lipid Bilayers*. CRC Press, Boca Raton, Florida (1995).

Durairaj, G. and I. Vijayakumar. Temperature-acclimation and phospholipid phase-transition in hypothalmic membrane phospholipids of garden lizards *Calotes versicolor*. Biochim. Biophys. Acta. **770**, 714–721 (1984).

Fragneto, G., T. Charitat, F. Graner, K. Mecke, L. Perino-Gallice, and E. Bellet-Amalric. A fluid floating bilayer. Europhys. Lett. **53**, 100–106 (2001).

Hønger, T., K. Jørgensen, R. L. Biltonen, and O. G. Mouritsen. Systematic relationship between phospholipase A_2 activity and dynamic lipid bilayer microheterogeneity. Biochemistry **28**, 9003–9006 (1996).

Hørup, P., K. Jørgensen, and O. G. Mouritsen. Phospholipase A_2 – an enzyme that is sensitive to the physics of its substrate. Europhys. Lett. **57**, 464–470 (2002).

Lemmich, J., K. Mortensen, J. H. Ipsen, T. Hønger, R. Bauer, and O. G. Mouritsen. Small-angle neutron scattering from multilamellar lipid bilayers: theory, model, and experiment. Phys. Rev. E **53**, 5169–5180 (1996).

Lipowsky, R. Generic interactions of flexible membranes. In *Handbook of Biological Physics* (A. J. Hoff, series ed.) *Vol. I: Structure and Dynamics of Membranes* (R. Lipowsky and E. Sackmann, eds.). Elsevier Science B. V., Amsterdam (1995) pp. 521–602.

Nagle, J. F. and S. Tristram-Nagle. Structure of lipid bilayers. Biochim. Biophys. Acta **1469**, 159–195 (2000).

Needham, D. and R. S. Nunn. Elastic deformation and failure of lipid bilayer membranes containing cholesterol. Biophys. J. **58**, 997–1009 (1990).

Trandum, C., P. Westh, K. Jørgensen and O.G. Mouritsen. A thermodynamic study of the effects of cholesterol on the interaction between liposomes and ethanol Biophys. J. **78**, 2486–2492 (2000).

13 PROTEINS AT LIPID MATTRESSES

Kinnunen, P. K. J. On the principles of functional ordering in biological membranes. Chem. Phys. Lipids **57**, 375–399 (1991).

de Kruijff, B. (ed.). *Lipid-Protein Interactions* Biochim. Biophys. Acta (Special Issue) **1376**, 245–484 (1998).

Heimburg, T. and R. L. Biltonen. The thermotropic behavior of dimyristoyl phosphatidylglycerol and its interaction with cytochrome c. Biochemistry **33**, 9477–9488 (1994).

Hong, H. and L. K. Tamm. Elastic coupling of integral membrane protein stability to lipid bilayer forces. Prof. Natl. Acad. Sci. USA **101**, 4065–4070 (2004).

Known membrane protein structures,
see http://www.mpibp-frankfurt.mpg.de/michel/public/memprotstruct.html.

Kolbe, M., H. Besir, L.-O. Essen, and D. Oesterhelt. Structure of the light-driven chloride pump halorhodopsin at 1.8 Å resolution. Science **288**, 1390–1396 (2000).

Merkel, R., P. Nassoy, A. Leung, K. Richie, and E. Evans. Energy landscapes of receptor-ligand bonds explored with dynamic force microscopy. Nature **397**, 50–53 (1999).

Mouritsen, O. G. and M. M. Sperotto. Thermodynamics of lipid protein-interactions in lipid membranes: the hydrophobic matching condition. In *Thermodynamics of Membrane Receptors and Channels*. (M. B. Jackson, ed.). CRC Press, Boca Raton, Florida (1993).

Murray D., A. Arbuzova, B. Honig, and S. McLaughlin. The role of electrostatic and non-polar interactions in the association of peripheral proteins with membranes. Curr. Topics Membr. **52**, 271–302 (2002).

Nielsen, L. K. *Small Scale Lateral Organisation in Lipid Membranes. An Atomic-force Microscopy Study.* PhD Thesis, Technical University of Denmark (2000).

Oesterhelt, F., D. Oesterhelt, M. Pfeiffer, A. Engel, H. E. Gaub, D. J. Müller. Unfolding pathways of individual bacteriorhodopsins. Science **288**, 143–146 (2000).

Palczewski, K., T. Kumasaka, T. Hori, C. A. Behnke, H. Motoshima, B. A. Fox, I. Le Trong, D. C. Teller, T. Okada, R. E. Stenkamp, M. Yamamoto, and M. Miyano. Crystal structure of rhodopsin: a G protein-coupled receptor. Science **289**, 739–745 (2000).

Ridder, A. N. J. A., W. van de Hoef, J. Stam, A. Kuhn, B. de Kruijff, and J. A. Killian. Importance of hydrophobic matching for spontaneous insertion of a single-spanning membrane protein. Biochemistry **41**, 4946–4952 (2002).

Rief, M. and H. Grubmüller. Force spectroscopy of single biomolecules. Chem. Phys. Chem. **3**, 255–261 (2002).

Scheuring, S., D. Fotiadis, C. Möller, S. A. Müller, A. Engel, and D. J. Müller. Single proteins observed by atomic force microscopy. Single Mol. **2**, 59–67 (2001).

Tuominen, E. K. J., C. J. A. Wallace, and P. K. J. Kinnunen. Phospholipid-cytochrome c interaction. Evidence for the extended lipid anchorage. J. Biol. Chem. **277**, 8822–8826 (2002).

von Heijne, G. (ed.). *Membrane Protein Assembly*. Springer-Verlag, Heidelberg (1997).

Watts, A. (ed.). *Protein-Lipid Interactions*. Elsevier, Amsterdam (1993).

White, S. H. and W. C. Wimley. Membrane protein folding and stability: physical principles. Annu. Rev. Biophys. Biomol. Struct. **28**, 319–365 (1999).

14 CHOLESTEROL ON THE SCENE

Bloom, M. and O. G. Mouritsen. The evolution of membranes. In *Handbook of Biological Physics* (A. J. Hoff, series ed.) *Vol. I: Structure and Dynamics of Membranes* (R. Lipowsky and E. Sackmann, eds.). Elsevier Science B. V., Amsterdam (1995) pp. 65–95.

Bloch, K. Sterol structure and membrane function. CRC Crit. Rev. **14**, 47–92 (1983).

Bloch, K. *Blondes in Venetian Paintings, the Nine-Banded Armadillo, and Other Essays in Biochemistry*. Yale University Press, New Haven (1994).

Cavalier-Smith, T. The origin of eukaryote and archaebacterial cells. Ann. NY Acad. Sci. **503**, 17–54 (1987).

Finegold, L.X. (ed.). *Cholesterol and Membrane Models*. CRC Press, Inc., Boca Raton, Florida (1993) pp. 223–257.

Goldstein, J. L. and M. S. Brown. The cholesterol quartet. Science **292**, 1310–1312 (2001).

Knoll, A. H. End of proterozoic eon. Sci. Amer. Oct. Issue, 64–73 (1991).

Knoll, A. H. The early evolution of eukaryotes. A geological perspective. Science **256**, 622–627 (1992).

Margulis, L. and D. Sagan. *Microcosmos – Four Billion Years of Evolution from Our Microbial Ancestors*. University of California Press, Berkeley (1997).

McConnell, H. M. and A. Radhakrishnan. Condensed complexes of cholesterol and phospholipids. Biochim. Biophys. Acta **1610**, 159–173 (2003).

McNamara, D. J. Dietary cholesterol and atherosclerosis. Biochim. Biophys. Acta **1529**, 310–320 (2000).

Miao, L., M. Nielsen, J. Thewalt, J. H. Ipsen, M. Bloom, M. J. Zuckermann, and O. G. Mouritsen. From lanosterol to cholesterol: structural evolution and differential effects on lipid bilayers. Biophys. J. **82**, 1429–1444 (2002).

Rohrer, M., P. Bouvier, and G. Ourisson. Molecular evolution of biomembranes: structural equivalents and phylogenetic precursors to cholesterol. Proc. Natl. Acad. Sci. USA **76**, 847–851 (1979).

Taubes, G. The soft science of dietary fats. Science **292**, 2536–2545 (2001).

Vance, D. E. and H. Van den Bosch (eds.). *Cholesterol in the Year 2000*. Biochim. Biophys. Acta **1529**, 1–373 (2000).

Xu, X. and E. London. The effect of sterol on membrane lipid domains reveal how cholesterol can induce lipid domain formation. Biochemistry **39**, 843–849 (2000).

Yeagle, P. L. (ed.). *Biology of Cholesterol*. CRC Press, Boca Raton, Florida (1988).

15 LIPIDS IN CHARGE

Andersen, O. S., C. Nielsen, A. M. Maer, J. A. Lundbaek, M. Goulian, and R. E. Koeppe II. Ion channels as tools to monitor lipid bilayer-membrane protein interactions: gramicidin channels as molecular force transducers. Meth. Enzymol. **294**, 208–224 (1998).

Attard, G. S., R. H. Templer, W. S. Smith, A. H. Hunt, and S. Jackowski. Modulation of CTP: phosphocholine cytidylyltransferase by membrane curvature elastic stress. Proc. Natl. Acad. Sci. USA **97**, 9032–9036 (2000).

Bezrukov, S. M. Functional consequences of lipid packing stress. Curr. Opin. Colloid Int. Sci. **5**, 237–243 (2000).

Bonifacino, J. S. and B. S. Glick. The mechanisms of vesicle budding and fusion. Cell **116**, 153–166 (2004).

Munro, S. Localization of proteins to the Golgi apparatus. Trends Cell. Biol. **8**, 11–15 (1998).

Brown, M. F. Influence of non-lamellar-forming lipids on rhodopsin. Curr. Topics Membr. **44**, 285–356 (1997).

Cantor, R. S. The influence of membrane lateral pressures on simple geometric models of protein conformational equilibria. Chem. Phys. Lipids **101**, 45–56 (1999).

Chernomordik, L. Non-bilayer lipids and biological fusion intermediates. Chem. Phys. Lipids **81**, 203–213 (1996).

Cornelius, F. Modulation of Na,K-ATPase and Na-ATPase activity by phospholipids and cholesterol. I. Steady-state kinetics. Biochemistry **40**, 8842–8851 (2001).

de Kruijff, B. Lipids beyond the bilayer. Nature **386**, 129–130 (1997).

Dumas, F., M. C. Lebrun, and J.-F. Tocanne. Is the protein/lipid hydrophobic matching principle relevant to membrane organization and function? FEBS Lett. **458**, 271–277 (1999).

Dumas, F., M. M. Sperotto, J.-F. Tocanne, and O. G. Mouritsen. Molecular sorting of lipids by bacteriorhodopsin in DMPC-DSPC lipid bilayers. Biophys. J. **73**, 1940–1953 (1997).

Epand, R. Lipid polymorphism and protein-lipid interactions. Biochim. Biophys. Acta **1376**, 353–368 (1998).

Gil, T., J. H. Ipsen, O. G. Mouritsen, M. C. Sabra, M. M. Sperotto, and M. J. Zuckermann. Theoretical analysis of protein organization in lipid membranes. Biochim. Biophys. Acta **1376**, 245–266 (1998).

Gohon, Y. and J.-L. Popot. Membrane protein-surfactant complexes. Curr. Opin. Colloid Interface Sci. **8**, 15–22 (2003).

Gruner, S. Lipid membrane curvature elasticity and protein function. In *Biologically Inspired Physics* (L. Peliti, ed.). Plenum Press, New York (1991) pp. 127–135.

Gullingsrud, J. and K. Schulten. Lipid bilayer pressure profiles and mechanosensitive channel gating. Biophys. J. **86**, 2883–2895 (2004).

Hinderliter, A., A. R. G. Dibble, R. L. Biltonen, and J. J. Sando. Activation of protein kinase C by coexisting di-acylglycerol-enriched and di-acylglycerol-poor lipid domains. Biochemistry **36**, 6141–6148 (1996).

Hinderliter, A., P. F. F. Almeida, C. E. Creutz, and R. L. Biltonen. Domain formation in a fluid mixed lipid bilayer modulated through binding of the C2 protein motif. Biochemistry **40**, 4181–4191 (2001).

Hunte, C. and H. Michel. Membrane protein crystallization. In *Membrane Protein Purification and Crystallization* (C. Hunte, G. von Jagow, and H. Schägger, eds.). Elsevier Science, New York (2003) pp. 143–160.

Jahn, R. and H. Grubmüller. Membrane fusion. Curr. Opin. Cell Biol. **14**, 488–495 (2002).

Jensen, M. Ø. *Molecular Dynamics Simulations of Proteins, Biomembrane Systems, and Interfaces*. PhD Thesis, Technical University of Denmark (2001).

Killian, J. A. and G. von Heijne. How proteins adapt to a membrane-water interface. Trends Biochem. Sci. **25**, 429–434 (2000).

Kinnunen, P. K. J. (ed.). *Peripheral Interactions on Lipid Surfaces: Towards a New Biomembrane Model.* Chem. Phys. Lipids (Special Issue) **101**, 1–137 (1999).

Lee, A. G. Lipid-protein interactions in biological membranes: a structural perspective. Biochim. Biophys. Acta **1612**, 1–40 (2003).

Lundbaek, J. A., P. Birn, A. J. Hansen, R. Søgaard, C. Nielsen, J. Girshman, M. J. Bruno, S. E. Tape, J. Egebjerg, D. V. Greathouse, G. L. Mattice, R. E. Koeppe II, and O. S. Andersen, Regulation of sodium channel function by bilayer elasticity – the importance of hydrophobic coupling. J. Gen. Physiol. **121** 599–621 (2004).

Mitra, K., I. Ubarretxena-Belandia, T. Taguchi, G. Warren, and D. M. Engelman. Modulation of the bilayer thickness of exocytic pathway membranes by membrane proteins rather that cholesterol. Proc. Natl. Acad. Sci. USA **101**, 4083–4088 (2004).

Mouritsen, O. G. Self-assembly and organization of lipid-protein membranes. Curr. Opin. Colloid Interface Sci. **3**, 78–87 (1998).

Mouritsen, O. G. and M. Bloom. Mattress model of lipid-protein interactions in membranes. Biophys. J. **46**, 141–153 (1984).

Nielsen, C., M. Goulian, and O. S. Andersen. Energetics of inclusion-induced bilayer deformations. Biophys. J. **74**, 1966–1983 (1998).

Niu, S.-L., D. C. Mitchell, S. Y. Lim, Z.-M. Wen, H.-Y. Kim, N. Salem, Jr., and B. J. Litman. Reduced G protein-coupled signaling efficiency in retinal rod outer segments in response to n-3 fatty acid deficiency. J. Biol. Chem. **279**, 31098–31104 (2004).

Noguchi, H. and M. Takasu. Fusion pathways of vesicles: a Brownian dynamics simulation. J. Chem. Phys. **115**, 9547–9551 (2001).

Nollert, P. J. Navarro and E. M. Landau. Crystallization of membrane proteins in cubo. Meth. Enzymol. **343**, 183–199 (2002).

Oliver, D., C.-C. Lien, M. Soom, T. Baukrowitz, P. Jonas, and B. Fakler. Functional conversion between A-type and delayed rectifier K^+ channels by membrane lipids. Science **304**, 265–270 (2004).

Perozo, E., A. Kloda, D. M. Cortes, and B. Martinac. Physical principles underlying the transduction of bilayer deformation forces during mechanosensitive channel gating. Nature Struct. Biol. **9**, 696–703 (2002).

Sprong, H., P. van der Sluijs, and G. van Meer. How proteins move lipids and lipids move proteins. Nature Rev. Mol. Cell Biol. **2**, 504–513 (2001).

Tamm, L. K., X. Han, Y. Li, and A. L. Lai. Structure and function of membrane fusion peptides. Biopolymers (Peptide Science) **66**, 249–260 (2002).

Weiss, M. and T. Nilsson. In a mirror dimly: tracing the movements of molecules in living cells. Trends Cell Biol. **14**, 267–273 (2004).

Weiss, T.M., P. C. A. Van der Wel, J. A. Killian, R. E. Koeppe II, and H. W. Huang. Hydrophobic mismatch between helices and lipid bilayers. Biophys. J. **84**, 379–385 (2003).

16 BEING SMART – A FISHY MATTER OF FACT

Bloom, M. Evolution of membranes from a physics perspective. In *In Search of a New Biomembrane Model.* (O. G. Mouritsen and O. S. Andersen, eds.). Biol. Skr. Dan. Vid. Selsk. **49**, 13–17 (1998).

Broadhurst, C. L., S. C. Cunnane, and M. A. Crawford. Rift Valley lake fish and shellfish provided brain-specific nutrition for early homo. Brit. J. Nutrition **79**, 3–21 (1998).

Broadhurst, C. L., Y. Wang, M. A. Crawford, S. C. Cunnane, J. E. Parkington, and W. E. Schmid. Brain-specific lipids from marine, lacustrine, or terrestrial food resources: potential impact on early African Homo sapiens. Comp. Biochem. Physiol. B: Biochem. Mol. Biol. **131**, 653–673 (2002).

Christensen, C. and E. Christensen. Fat consumption and schizophrenia. Acta Psychiatr. Scand. **78**, 587–591 (1988).

Crawford, M. and D. Marsh. *The Driving Force.* Harper and Row, Publ., New York (1989).

Crawford, M. A., M. Bloom, C. L. Broadhurst, W. F. Schmidt, S. C. Cunnane, C. Galli, K. Gehbremeskel, F. Linseisen, J. Lloyd-Smith, and J. Parkington. Evidence for the unique function of DHA during the evolution of the modern hominid brain. Lipids **34**, S39–S47 (1999).

Gawrisch, K., N. V. Eldho, and L. L. Holte. The structure of DHA in phospholipid membranes. Lipids **38**, 445–452 (2003).

Gurr, M. L. *Lipids in Nutrition and Health: A Reappraisal.* The Oily Press, Bridgwater (1999).

Horrobin, D. F. Schizophrenia: the illness that made us human. Medical Hypotheses **50**, 269–288 (1998).

Horrobin, D. F. *The Madness of Adam and Eve.* Bantam Press, London (2001).

Lauritsen, L., H. S. Hansen, M. H. Jørgensen, and K. F. Michaelsen. The essentiality of long chain n-3 fatty acids in relation to development and function of the brain and retina. Prog. Lip. Res. **40**, 1–94 (2001).

Pond, C. M. *The Fats of Life.* Cambridge University Press, Cambridge (1998).

17 LIQUOR AND DRUGS – AS A MATTER OF FAT

Bayley, H. Building doors into cells. Sci. Amer. **277** (3), 42–47 (1997).

Bechinger, B. Structure and function of channel-forming peptides: magainins, cecropins, melittin, and alamethicin. J. Memb. Biol. **156**, 197–211 (1997).

Boman, H. G., J. Marsh, and J. A. Goode. *Antimicrobial Peptides.* John Wiley and Sons, Chichester (1994).

Cantor, R. S. The lateral pressure profile in membranes: a physical mechanism of general anesthesia. Biochemistry **36**, 2339–2344 (1997).

Deamer, D. W., A. Kleinzeller, and D. M. Fambrough (eds.). *Membrane Permeability. 100 Years Since Ernest Overton.* Academic Press, London (1999).

Dietrich, R. A., T. V. Dunwiddie, R. A. Harris, and V. G. Erwin. Mechanism of action of ethanol: initial central nervous system actions. Pharmacol. Rev. **41**, 489–527 (1989).

Fenster, J. M. *Ether Day.* Harper Collins Publ., New York (2001).

Gruner, S. M. Intrinsic curvature hypothesis for biomembrane lipid composition: a role for non-bilayer lipids. Proc. Natl. Acad. Sci. USA **82**, 3665–3669 (1985).

Lohner, K., and R. M. Epand. Membrane interactions of hemolytic and antimicrobial peptides. Adv. Biophys. Chem. **6**, 53–66 (1997).

Lundbaek, J. A. and O. S. Andersen. Lysophospholipids modulate channel function by altering the mechanical properties of lipid bilayers. J. Gen. Physiol. **104**, 645–673 (1994).

Lundbaek, J. A., A. M. Maer, and O. S. Andersen. Lipid bilayer electrostatic energy, curvature stress and assembly of gramicidin channels. Biochemistry **36**, 5695–5701 (1997).

Matsuzaki, K. Molecular mechanisms of membrane perturbations by anti-microbial peptides. In *Development of Novel Anti-microbial Agents: Emerging Strategies* (K. Lohner, ed.). Horizon Scientific Press, Wymondham, UK (2001) pp. 167–181.

Matsuzaki, K. Why and how are peptide-lipid interactions utilized for self defense? Biochem. Soc. Trans. **29**, 598–601 (2001).

Mihic, S., Q. Ye, M. J. Wick, V. V. Koltchine, M. D. Krasowski, S. E. Finn, M. P. Mascia, C. F. Valenzuela, K. M. Hanson, E. P. Greenblatt, R. A. Harris, and N. L. Harrison. Sites of alcohol and volatile anesthetic action on $GABA_A$ and glycine receptors, Nature **389**, 385–389 (1997).

Mouritsen, O. G. and K. Jørgensen. A new look at lipid-membrane structure in relation to drug research. Pharm. Res. **15**, 1507–1519 (1998).

Franks, N. P. and W. R. Lieb. Where do general anesthetics act? Nature **274**, 339–342 (1978).

Roth, S. H. and K. W. Miller (eds.). *Molecular and Cellular Mechanisms of Anesthetics.* Plenum Publishing Corporation, New York (1986).

Schreier, S., S. V. P. Malheiros, and E. de Paula. Surface active drugs: self-association and interaction with membranes and surfactants. Physico-chemical and biological aspects. Biochim. Biophys. Acta **1508**, 210–234 (2000).

Testa, B., H. van de Waterbeemd, G. Folkers, and R. Guy (eds.). *Pharmacokinetic Optimization in Drug Research: Biological, Physicochemical and Computational Strategies.* Wiley-Verlag Helvetica Chemica Acta, Zürich (2000).

Wallace, B. A. Recent advances in the high resolution structure of bacterial channels: gramicidin A. J. Struct. Biol. **121**, 123–141 (1998).

Wiedemann, I., E. Breukink, C. Van Kraaij, O. P. Kuipers, G. Bierbaum, B. de Kruijff, and H.-G. Sahl. Specific binding of nisin to the peptidoglycan precursor Lipid II combines pore formation and inhibition of cell wall biosynthesis for potent antibiotic activity. J. Biol. Chem. **276**, 1772–1779 (2001).

18 LIPID EATERS

Berg, O. G., M. H. Gelb, M.-D. Tsai, and M. K. Jain. Interfacial enzymology: the secreted phospholipase A_2 paradigm. Chem. Rev. **101**, 2613–2653 (2001).

Buckland, W. R. and D. C. Wilton. The antibacterial properties of secreted phospholipases A_2. Biochim. Biophys. Acta **1488**, 71–82 (2000).

Colton, C. (ed.). *Reactive Oxygen Species in Biological Systems*. Kluwer Academic/Plenum Publ., New York (1999).

Goñi, F. and A. Alonso. Sphingomyelinases: enzymology and membrane activity. FEBS Lett. **531**, 38–46 (2002).

Grainger, D. W., A. Reichert, H. Ringsdorf, and C. Salesse. An enzyme caught in action: direct imaging of hydrolytic function and domain formation of phospholipase A_2 in phosphatidylcholine monolayers. FEBS Lett. **252**, 72–82 (1989).

Gurr, M. L. *Lipids in Nutrition and Health: A Reappraisal*. The Oily Press, Bridgwater (1999).

Hønger, T., K. Jørgensen, D. Stokes, R. L. Biltonen, and O. G. Mouritsen. Phospholipase A_2 activity and physical properties of lipid-bilayer substrates. Meth. Enzymol. **286**, 168–190 (1997).

Kaasgaard, T., J. H. Ipsen, O. G. Mouritsen, and K. Jøgensen. In situ atomic force microscope imaging of phospholipase A_2 lipid bilayer hydrolysis. J. Probe Microscopy **2**, 169–175 (2001)

Nielsen, L. K., J. Risbo, T. H. Callisen, and T. Bjørholm. Lag-burst kinetics in phospholipase A_2 hydrolysis of DPPC bilayers visualized by atomic force microscopy. Biochim. Biophys. Acta **1420**, 266–271 (1999).

Nurminen, T. A., J. M. Holopainen, H. Zhao, and P. K. J. Kinnunen. Observation of topical catalysis by sphingomyelinase coupled to microspheres. J. Amer. Chem. Soc. **124**, 12129–12134 (2002).

Pond, C. M. *The Fats of Life*. Cambridge University Press, Cambridge (1998).

Rubin, B. and E. A. Dennis. *Lipases: Characterization and Utilization*. Academic Press, Orlando (1997).

Sanchez, S. A., L. A. Bagatolli, E. Gratton, and T. L. Hazlett. A two-photon view of an enzyme at work: *Crotalus atrox* venom PLA2 interaction with single-lipid and mixed-lipid giant uni-lamellar vesicles. Biophys. J. **82**, 2232–2243 (2002).

Six, D. A. and E. A. Dennis. The expanding superfamily of phospholipase A_2 enzymes: classification and characterization. Biochim. Biophys. Acta **1488**, 1–19 (2000).

Zhou, F. and K. Schulten. Molecular dynamics study of the activation of phospholipase A_2 on a membrane surface. Proteins: Structure, Function, and Genetics **25**, 12–27 (1996).

19 POWERFUL AND STRANGE LIPIDS AT WORK

Bazan, N. G. and R. J. Flower. Lipids in pain control. Nature **420**, 135–138 (2002).

Balny, C., P. Mansson, and K. Heermans (eds.). *Frontiers in High Pressure Biochemistry and Biophysics*. Biochim. Biophys. Acta (Special Issue) **1595**, 1–399 (2002).

Bittar, E. E. and N. Bittar. (eds.). *Membranes and Cell Signaling*. Jai Press, Greenwich, Connecticut (1997).

Cravatt, B. F., O. Prospero-Garcia, G. Siuzdak, N. B. Gilula, S. J. Henriksen, D. L. Boger, and R. A. Lerner. Chemical characterization of a family of brain lipids that induce sleep. Science **268**, 1506–1509 (1995).

Dennis, E. A., I. Varela-Nieto, and A. Alonso (eds.). *Lipid Signalling: Cellular Events and Their Biophysical Mechanisms.* FEBS Lett. **531**, 1–109 (2002).

Elias, P. M. Epidermal barrier function: intercellular epidermal lipid structures, origin, composition and metabolism. J. Control. Release **15**, 199–208 (1991).

Fiers, W., R. Beyaert, W. Declercq, and P. Vandenabeele. More than one way to die: apoptosis, necrosis and reactive oxygen damage. Oncogene **18**, 7719–7730 (1999).

Forslind, B. A domain mosaic model of the skin barrier. Acta. Derm. Venerol. **74**, 1–6 (1994).

Holopainen, J. M., M. I. Angelova, and P. K. J. Kinnunen. Budding of vesicles by asymmetrical enzymatic formation of ceramide in giant liposomes. Biophys. J. **78**, 830–838 (2000).

Howland, J. L. *The Surprising Archaea. Discovering Another Domain of Life.* Oxford University Press, Oxford (2000).

Kinnunen, P. K. J. On the molecular-level mechanisms of peripheral protein-membrane interactions induced by lipids forming inverted non-lamellar phases. Chem. Phys. Lipids **81**, 151–166 (1996).

Kinnunen, P. K. J., J. M. Holopainen, and M. I. Angelova. Giant liposomes as model biomembranes for roles of lipids in cellular signaling. In *Giant Vesicles* (Luisi, P. L. and P. Walde, eds.). John Wiley and Sons Ltd., New York (2000).

Kolesnick, R. N. Sphingomyelin and derivatives as cellular signals. Prog. Lipid Res. **30**, 1–38 (1991).

Krönke, M. Biophysics of ceramide signaling: interaction with proteins and phase transition of membranes. Chem. Phys. Lipids **101**, 109–121 (1999).

Landman, L. Epidermal permeability barrier: transformation of lamellar granule-disks into intercellular sheets by a membrane-fusion process, a freeze-fracture study. J. Invest. Dermatol. **87**, 202–209 (1986).

Levade, T., Y. Hannun, and S. Spiegel (eds.). *Lipids in Apoptosis.* Biochim. Biophys. Acta **1585**, 51–221 (2002).

Nikaido, H. Permeability of the lipid domain of bacterial membranes. In *Membrane Transport and Information Storage* (R. C. Aloia, C. Curtain, and L. M. Gordon, eds.). Wiley-Liss, Inc., New York (1990) pp. 165–190.

Norlén, L. Skin barrier formation: the membrane folding model. J. Invest. Dermatol. **117**, 823–829 (2001).

Norlén, L. Skin barrier structure and function: the single gel phase model. J. Invest. Dermatol. **117**, 830–836 (2001).

Nurminen, T. A., J. M. Holopainen, H. Zhao, and P. K. J. Kinnunen. Observation of topical catalysis by sphingomyelinase coupled to microspheres. J. Amer. Chem. Soc. **124**, 12129–12134 (2002).

Ourisson, G. Biomembranes, a more general perspective. In *Biomembranes and Nutrition.* (C. L. Léger and G. Béréziat, eds.). Colloque INSERM **195**, 21–36 (1989).

Sen, A., M. E. Daly, and S. W. Hui. Transdermal insulin delivery using enhanced electroporation. Biochim. Biophys. Acta **1564**, 5–8 (2002).

Silvius, J. Lipid modifications of intracellular signal-transducing proteins. J. Liposome Res. **9**, 1–19 (1999).

Sparr, E. and H. Wennerström. Responding phospholipid membranes – interplay between hydration and permeability. Biophys. J. **81**, 1014–1028 (2001).

Tigyi, G. and E. J. Goetzl (eds.). *Lysolipid Mediators in Cell Signaling and Disease*. Biochim. Biophys. Acta (Special Issue) **1582**, 1–317 (2002).

Yayanos, A. A. The properties of deep-sea piezophilic bacteria and their possible use in biotechnology. In *Advances in High Pressure Bioscience and Biotechnology* (H. Ludwig, ed.). Springer-Verlag, Berlin (1999) pp. 3–9.

20 SURVIVAL BY LIPIDS

Allen, T. M. and P. R. Cullis. Drug delivery systems: entering the mainstream. Science **303**, 1818–1822 (2004).

Bangham, A. D. Surrogate cells or Trojan horses. The discovery of liposomes. Bioessays **17**, 1081–1088 (1995).

Barenholtz, Y. Liposome application: problems and prospects. Curr. Opin. Colloid. Interface Sci. **6**, 66–77 (2001).

Chapman, D. Biomembranes and new hemocompatible materials. Langmuir **9**, 39–45 (1993).

Cevč, G. Material transport across permeability barriers by means of lipid vesicles. In *Handbook of Biological Physics: Vol. 1* (R. Lipowsky and E. Sackmann, eds.). Elsevier Science, Amsterdam (1995) pp. 465–490.

Cornell, B. A., V. L. B. Braach-Maksvytis, L. G. King, P. D. J. Osman, B. Raguse, L. Wieczorek, and J. L. Pace. A biosensor that uses ion-channel switches. Nature **387**, 580–583 (1997).

Cullis, P. R. Commentary: liposomes by accident. L. Liposome Res. **10**, ix–xxiv (2000).

Davidsen, J., K. Jørgensen, T. L. Andresen, and O. G. Mouritsen. Secreted phospholipase A2 as a new enzymatic trigger mechanism for localized liposomal drug release and adsorption in diseased tissue. Biochim. Biophys. Acta **1609**, 95–101 (2003).

Evans, E., H. Bowman, A. Leung, D. Needham, and D. Tirrell. Biomembrane templates for nanoscale conduits and networks. Science **272**, 933–935 (1996).

Glebart, W. M., R. F. Bruinsma, P. A. Pincus, and V. A. Parsegian. DNA-inspired electrostatics. Physics Today, September, 38–44 (2000).

Godsell, D. A. *Bionanotechnology*. Wiley-Liss, Inc., Hoboken, New Jersey (2004).

Gregoriadis, G., E. J. Wills, C. P. Swain, and A. S. Tavill. Drug-carrier potential of liposomes in cancer chemotherapy. Lancet 1, 1313–1316 (1974).

Janoff, A.S. *Liposomes – Rational Design*. Marcel Dekker Inc., New York (1999).

Jørgensen, K., J. Davidsen, and O. G. Mouritsen. Biophysical mechanisms of phospholipase A_2 activation and their use in liposome-based drug delivery. FEBS Lett. **531**, 23–27 (2002).

Karlsson, A., R. Karlsson, M. Karlsson, A.-S. Cans, A. Strömberg, F. Ryttsén, and O. Orwar. Networks of nanotubes and containers. Nature **409**, 150–152 (2001).

Kim, D. H. and D. Needham. Lipid bilayers and monolayers: characterization using micropipette manipulation techniques. In *Encyclopedia of Surface and Colloid Science* (A. Hobbard, ed.). Marcel Dekker, New York (2002) pp. 3057–3086.

Lasic, D. D. (ed.). *Liposomes in Gene Delivery*. CRC Press, Boca Raton, Florida (1997).

Lasic, D. D. and F. Martin. (eds.). *Stealth Liposomes*. CRC Press, Boca Raton, Florida (1995).

Lasic, D. D. and D. Needham. The "Stealth" liposome: a prototypical biomaterial. Chem. Rev. **95**, 2601–2628 (1995).

Lasic, D. D. and D. Papahadjopoulos (eds.). *Medical Applications of Liposomes*. Elsevier, Amsterdam (1998).

Larsson, K. *Lipids – Molecular Organization, Physical Functions, and Technical Applications*. The Oily Press, Dundee (1994).

Mann, J. *The Elusive Magic Bullet. The Search for the Perfect Drug*. Oxford University Press, Oxford (1999).

Needham, D., G. Anyarambhatla, G. Kong, and M. W. Dewhirst. A new temperature-sensitive liposome for use with mild hyperthermia: characterization and testing in a human tumor xenograft model. Cancer Res. **60**, 1197–1201 (2000).

Peet, M., I. Glen, and D. F. Horrobin (eds.). *Phospholipid Spectrum Disorder in Psychiatry*. Marius Press, Carnforth, UK (1999).

Ratner, M. and D. Ratner. *Nanotechnology*. Prentice Hall, Upple Saddle River, New Jersey (2003).

Rubahn, H.-G. *Nanophysik und Nanotechnologie*. 2nd ed., Teubner, Stuttgart (2004).

Sackmann, E. Supported membranes: scientific and practical applications. Science **271**, 43–48 (1996).

Sackmann, E. and M. Tanaka. Supported membranes on soft polymer cushions: fabrication, characterization, and applications. Trends Biotech. **18**, 58–64 (2000).

Schliwa, M. (ed.). *Molecular Motors*. John Wiley and Sons, New York (2003).

Schnur, J. M. Lipid tubules: a paradigm for molecularly engineered structures. Science **262**, 1669–1676 (1993).

Shimomura, M. and T. Sawadaishi. Bottom-up strategy of materials fabrication: a new trend in nano-technology of soft materials. Curr. Opin. Colloid Int. Sci. **6**, 11–16 (2001).

van Oudenaarden, A. and S. G. Boxer. Brownian ratchets: molecular separations in lipid bilayers supported on patterned arrays. Science **285**, 1046–1048 (1999).

Vilstrup, P. (ed.). *Microencapsulation of Food Ingredients*. Leatherhead Publ., Leatherhead, UK (2001).

Whitesides, G. M. and B. Grzybowski. Self-assembly at all scales. Science **295**, 2418–2421 (2002).

EPILOGUE: FAT FOR FUTURE

Ainsworth, C. Love that fat. New Scientist **2256**, 37–39 (2000).

Ewin, J. *Fine Wines and Fish Oil: the Life of Hugh MacDonald Sinclair*. Oxford University Press, Oxford (2002).

Clarke, S. D. The multi-dimensional regulation of gene expression by fatty acids: polyunsaturated fats as nutrient sensors. Curr. Opin. Lipidol. **15**, 13–18 (2004).

Crawford, M. and D. Marsh. *The Driving Force. Food, Evolution, and the Future*. Harper and Row Publ., New York (1989).

Editorial. Bigger isn't always better. Nature **418**, 353 (2002).

Enard, W., P. Khaitovich, J. Klose, S. Zöllner, F. Heissig, P. Giavalisco, K. Nieselt-Struwe, E. Muchmore, A. Varki, R. Ravid, G. M. Doxiadis, R. E. Bontrop, and S. Pääbo. Intra- and interspecific variation in primate gene expression patterns. Science **296**, 340–343 (2002).

Horrobin, D. F. Innovation in the pharmaceutical industry. J. Royal. Soc. Med. **93**, 341–345 (2000).

Peet, M., I. Glen, and D. F. Horrobin (eds.). *Phospholipid Spectrum Disorder in Psychiatry*. Marius Press, Carnforth, UK (1999).

Pond, C. *The Fats of Life*. Cambridge University Press, Cambridge (1998).

Sears, B. *The Omega Rx Zone*. ReganBooks, Harper Collins Publ., New York (2002).

Science Special Section on Obesity. Science **299**, 845–860 (2003).

Taubes, G. The soft science of dietary fats. Science **291**, 2536–2545 (2001).

Sources for Figures

The figures of the present book have been produced with the help of a number of colleagues who gave me their permission to use their artwork. Furthermore, many publishers have been so kind to grant me permission to reprint already published graphic work to which they hold the copyrights. I am indebted to Dr. Walter Shaw from Avanti Polar Lipids, Inc. for permission to use the Avanti structural formulas for lipids. The following list provides references to the sources I have used for the figures.

Fig. 1.1 (a) Adapted from http://www.bact.wisc.edu/Bact303/MajorGroupsOf Prokaryotes.

(b) Adapted from http://www.its.caltech.edu/boozer/symbols/e-coli.jpg.

(c) Adapted from http://web.ncifcrf.gov/rtp/ial/eml/rbc.asp.

Fig. 1.2 Adapted from Howland, J. L. *The Surprising Archae.* Oxford University Press, Oxford (2000).

Fig. 1.3 Adapted from http://www.nyu.edu/pages/mathmol/textbook/life.html.

Fig. 1.4 (a) Adapted from http://www.brooklyn.cuny.edu/bc/ahp/SDPS/graphics/Cellulose.GIF.

(b) Adapted from http://www.accelrys.com/cerius2/images/profiles_verify_thumb.gif.

(c) Adapted from http://www.genelex.com/paternitytesting/images/dna-molecule.jpg.

Fig. 1.5 Reprinted from Gennis, R. B. *Biomembranes. Molecular Structure and Function.* Springer-Verlag, Berlin (1989). With permission from Springer-Verlag.

Fig. 1.6 Courtesy of Dr. Terry Beveridge.

Fig. 1.7 Illustration by Ove Broo Sørensen.

Fig. 3.5 (a) Adapted from Callisen, T. H. and Y. Talmon. Direct imaging by cryo- TEM shows membrane break-up by phospholipase A_2, enzymatic activity. Biochemistry **37**, 10987 10993 (1998). Courtesy of Dr. Thomas H. Callisen. (b) and (c) Courtesy of Dr. Jonas Henriksen. (d) Courtesy of Dr. Luis Bagatolli.

Fig. 3.6 (c) Adapted from http://www.chembio.uoguelph.ca/educmat/chm730/gif/porinr01.gif.

(d) Adapted from http://www.chembio.uoguelph.ca/educmat/chm730/gif/gfp.gif.

Fig. 3.7 Courtesy of Dr. Morten Ø. Jensen.

Fig. 3.8 Courtesy of Drs. Evan Evans and Dennis Kim.

Fig. 4.3 Courtesy of Jacob Sonne.

Fig. 4.4 Adapted from Jönsson, B., B. Lindman, K. Holmberg, and B. Kronberg. *Surfactants and Polymers in Aqueous Solution.* 2nd ed. John Wiley and Sons, New York (2002). Courtesy of Dr. Björn Lindman.

Fig. 4.6 Courtesy of Dr. Olaf Sparre Andersen.
Fig. 5.1 (a) and (c) Courtesy of Dr. Mohamed Laradji. (b) Courtesy of Dr. Gerhard Besold. (d) Adapted from Pieruschka, P. and S. Marcelja. Monte Carlo simulation of curvature-elastic interfaces. Langmuir **10**, 345–350 (1994). Courtesy of Dr. Stepjan Marcelja.
(e) Courtesy of Dr. Luis Bagatolli.
Fig. 5.2 (b) Courtesy of Dr. David Boal. (c) Courtesy of Dr. Aki Kusumi.
Fig. 5.4 (a) Courtesy of Drs. Thomas Kaasgaard and Anna Celli. (b) Adapted from http://www.carnegieinstitution.org/.../images/red.
Fig. 5.6 Courtesy of Dr. Jonas Henriksen.
Fig. 5.7 (a) Courtesy of Dr. Beate Klösgen.
Fig. 6.2 Adapted from Porte, G. From giant micelles to fluid membranes: polymorphism in dilute solutions of surfactant molecules. In *Soft Matter Physics* (M. Daoud and C. E. Williams, eds.). Springer-Verlag, Berlin (1995) pp. 155–185. With permission from Springer-Verlag.
Fig. 6.3 Courtesy of Dr. Hans-Günther Döbereiner.
Fig. 6.4 Courtesy of Dr. Hans-Günther Döbereiner.
Fig. 6.5 (a) and (e) Courtesy of Dr. Jonas Henriksen. (b)–(d) Courtesy of Dr. Hans-Günther Döbereiner.
Fig. 6.6 Courtesy of Dr. Hans-Günther Döbereiner.
Fig. 6.7 Adapted from Lim, G. H. W., M. Wortis, and R. Mukhopadhyay. Stomatocyte-discocyte-echinocyte sequence of the human red blood cell: evidence for the bilayer-couple hypothesis from membrane mechanics. Proc. Natl. Acad. Sci. **99**, 16766–16769 (2002).
Fig. 7.2 Adapted from Gennis, R. B. *Biomembranes. Molecular Structure and Function.* Springer-Verlag, Berlin (1989). With permission from Springer-Verlag.
Fig. 8.1 (a) Adapted from Tieleman, D. P., S. J. Marrink, H. J. C. Berendsen. A computer perspective of membranes: molecular dynamics studies of lipid bilayer systems. Biochim. Biophys. Acta **1331**, 235–270 (1997). Courtesy of Dr. Peter Tieleman. (b) Adapted from White, S. H. and M. C. Wiener. The liquid-crystallographic structure of fluid lipid bilayer membranes. In *Biological Membranes. A Molecular Perspective for Computation and Experiment* (K. M. Merz and B. Roux, eds.). Birkhäuser Publ. Co., New York (1996) pp. 127–144. Courtesy of Dr. Stephen White.
Fig. 8.3 Adapted from Lemmich, J., K. Mortensen, J. H. Ipsen, T. Hønger, R. Bauer, and O. G. Mouritsen. Small-angle neutron scattering from multilamellar lipid bilayers: theory, model, and experiment. Phys. Rev. E **53**, 5169–5180 (1996).
Fig. 8.5 Courtesy of Dr. Thomas Schmidt.
Fig. 8.6 Courtesy of Dr. Bruce Paul Gaber.
Fig. 9.2 Adapted from Bolhuis, P. *Liquid-like Behavior in Solids. Solid-like Behavior in Liquids.* PhD Thesis, University of Utrecht (1996).
Fig. 9.3 Courtesy of Dr. Thomas Kaasgaard.
Fig. 9.4 Courtesy of Dr. Morten Ø. Jensen.
Fig. 9.5 Courtesy of Dr. Olle Edholm.
Fig. 9.6 Courtesy of Dr. Thomas Kaasgaard.
Fig. 9.7 Adapted from Ipsen, J. H. and O. G. Mouritsen. Modeling the phase equilibria in two-component membranes of phospholipids with different acyl-chain lengths. Biochim. Biophys. Acta **944**, 121–134 (1988).

Fig. 9.8 Adapted from Ipsen, J. H. and O. G. Mouritsen. Modeling the phase equilibria in two-component membranes of phospholipids with different acyl-chain lengths. Biochim. Biophys. Acta. **944**, 121–134 (1988).
Fig. 9.9 Courtesy of Drs. Michael Patra and Ilpo Vattulainen.
Fig. 9.10 Courtesy of Drs. Morten Jensen and Martin J. Zuckermann.
Fig. 10.1 Courtesy of Dr. Lars Kildemark.
Fig. 10.2 Courtesy of Drs. Dennis Kim and David Needham.
Fig. 10.3 Courtesy of Dr. Lars Kildemark.
Fig. 10.4 Courtesy of Dr. Lars Kildemark.
Fig. 10.5 Courtesy of Drs. Harden McConnell and Helmuth Möhwald.
Fig. 10.6 Courtesy of Dr. Lars Kildemark.
Fig. 10.7 Courtesy of Dr. Lars Kildemark.
Fig. 10.8 Courtesy of Drs. Lars Kildemark and Thomas Bjørnholm.
Fig. 10.9 Adapted from http://classes.yale.edu/fractals/Panorama/Biology/Physiology/Physiology.html.
Fig. 10.10 Courtesy of Dr. Jesus Pérez-Gil.
Fig. 10.11 Courtesy of Drs. Adam C. Simonsen and Luis Bagatolli.
Fig. 11.1 Courtesy of Dr. Kent Jørgensen.
Fig. 11.2 Adapted from Leidy, C., W. F. Wolkers, O. G. Mouritsen, K. Jørgensen, and J. H. Crowe. Lateral organization and domain formation in a two-component lipid membrane system. Biophys. J. **80**, 1819–1828 (2001).
Fig. 11.3 Courtesy of Dr. Luis Bagatolli.
Fig. 11.5 Courtesy of Drs. Thomas Kaasgaard and Chad Leidy.
Fig. 11.6 Courtesy of Dr. Aki Kusumi. (http://www.supra.bio.nagoya-u.ac.jp/lab/slide2.html)
Fig. 11.7 Adapted from www.glycoforum.gr.jp/science/word/glycolipid/GLD01E.html.
Fig. 11.8 Adapted from Rinia, H. A., M. M. E. Snel, J. P. J. M. van der Eerden, and B. de Kruijff. Visualizing detergent resistant domains in model membranes with atomic force microscopy. FEBS Lett. **501**, 92–96 (2001). Courtesy of Dr. Ben de Kruijff.
Fig. 12.1 (a) Courtesy of Jesper Sparre Andersen. (b) Adapted from Trandum, C., P. Westh, K. Jørgensen, and O.G. Mouritsen. A thermodynamic study of the effects of cholesterol on the interaction between liposomes and ethanol. Biophys. J. **78**, 2486–2492 (2000).
Fig. 12.2 (a) Adapted from Corvera, E., O. G. Mouritsen, M. A. Singer, and M. J. Zuckermann. The permeability and the effects of acyl-chain length form phospholipid bilayers containing cholesterol: theory and experiments. Biochim. Biophys. Acta **1107**, 261–270 (1992). (b) Adapted from Trandum, C., P. Westh, K. Jørgensen, and O.G. Mouritsen. A thermodynamic study of the effects of cholesterol on the interaction between liposomes and ethanol. Biophys. J. **78**, 2486–2492 (2000).
Fig. 12.3 (a) Courtesy of Dr. Thomas Heimburg. (b) Adapted from Ipsen, J. H., K. Jørgensen, and O. G. Mouritsen. Density fluctuations in saturated phospholipid bilayers increase as the acyl–chain length decreases. Biophys. J. **58**, 1099–1107 (1990).
Fig. 12.4 (b) Adapted from Lemmich, J., K. Mortensen, J. H. Ipsen, T. Hønger, R. Bauer, and O. G. Mouritsen. Small-angle neutron scattering from multilamellar lipid bilayers: theory, model, and experiment. Phys. Rev. E **53**, 5169–5180 (1996).

Fig. 12.5 Courtesy of Dr. Kent Jørgensen.
Fig. 12.6 Adapted from Durairaj, G. and I. Vijayakumar. Temperature-acclimatization and phospholipid phase-transition in hypothalmic membrane phospholipids of garden lizards *Calotes versicolor*. Biochim. Biophys. Acta. **770**, 714–721 (1984).
Fig. 13.2 Courtesy of Dr. Lars Kildemark.
Fig. 13.3 Courtesy of Dr. Morten Ø. Jensen.
Fig. 13.5 Courtesy of Dr. Morten Ø. Jensen.
Fig. 13.6 Adapted from Oesterhelt, F., D. Oesterhelt, M. Pfeiffer, A. Engel, H. E. Gaub, and D. J. Müller. Unfolding pathways of individual bacteriorhodopsins. Science **288**, 143–146 (2000). Courtesy of Dr. Daniel Müller.
Fig. 13.7 Adapted from Hong, H. and L. K. Tamm. Elastic coupling of integral membrane protein stability to lipid bilayer forces. Prof. Natl. Acad. Sci. USA **101**, 4065–4070 (2004). Copyright 2004 National Academy of Sciences, U.S.A. Courtesy of Dr. Lukas Tamm.
Fig. 14.1 Adapted from Bloom, M. and O. G. Mouritsen. The evolution of membranes. In *Handbook of Biological Physics* (A. J. Hoff, series ed.) *Vol. I: Structure and Dynamics of Membranes* (R. Lipowsky and E. Sackmann, eds.). Elsevier Science B. V., Amsterdam (1995) pp. 65–95.
Fig. 14.2 Adapted from Bloch, K. *Blondes in Venetian Paintings, the Nine-Banded Armadillo, and Other Essays in Biochemistry.* Yale University Press, New Haven (1994).
Fig. 14.3 Adapted from Miao, L., M. Nielsen, J. Thewalt, J. H. Ipsen, M. Bloom, M. J. Zuckermann, and O. G. Mouritsen. From lanosterol to cholesterol: structural evolution and differential effects on lipid bilayers. Biophys. J. **82**, 1429–1444 (2002).
Fig. 14.4 Adapted from Miao, L., M. Nielsen, J. Thewalt, J. H. Ipsen, M. Bloom, M. J. Zuckermann, and O. G. Mouritsen. From lanosterol to cholesterol: structural evolution and differential effects on lipid bilayers. Biophys. J. **82**, 1429–1444 (2002).
Fig. 14.5 Adapted from Miao, L., M. Nielsen, J. Thewalt, J. H. Ipsen, M. Bloom, M. J. Zuckermann, and O. G. Mouritsen. From lanosterol to cholesterol: structural evolution and differential effects on lipid bilayers. Biophys. J. **82**, 1429–1444 (2002).
Fig. 15.1 Adapted from Mouritsen, O. G. and K. Jørgensen. A new look at lipid-membrane structure in relation to drug research. Pharm. Res. **15**, 1507–1519 (1998).
Fig. 15.2 Courtesy of Dr. Morten Ø. Jensen.
Fig. 15.3 Courtesy of Dr. Morten Ø. Jensen.
Fig. 15.4 Adapted from Gil, T., J. H. Ipsen, O. G. Mouritsen, M. C. Sabra, M. M. Sperotto, and M. J. Zuckermann. Theoretical analysis of protein organization in lipid membranes. Biochim. Biophys. Acta **1376**, 245–266 (1998).
Fig. 15.5 Adapted from Sternberg, B., A. Watts, and Z. Cejka. Lipid induced modulation of the protein packing in two-dimensional crystals of bacteriorhodopsin. J. Struct. Biology **110**, 196–204 (1993). Courtesy of Drs. Brigitte Sternberg and Tony Watts.
Fig. 15.6 Adapted from Sackmann, E. Physical basis for trigger processes and membrane structures. In *Biological Membranes Vol. 5* (D. Chapman, ed.). Academic Press, London (1984) pp. 105–143.

Fig. 15.7 (a) Adapted from Lee, A. G. Lipid-protein interactions in biological membranes: a structural perspective. Biochim. Biophys. Acta **1612**, 1–40 (2003). (b) Adapted from Cornelius, F. Modulation of Na,K-ATPase and Na-ATPase activity by phospholipids and cholesterol. I. Steady-state kinetics. Biochemistry **40**, 8842–8851 (2001).

Fig. 15.8 (a) Courtesy of Dr. Paavo K. J. Kinnunen. (b) Courtesy of Dr. Olaf Sparre Andersen.

Fig. 15.9 Courtesy of Dr. Robert Cantor.

Fig. 15.10 Adapted from Gullingsrud, J., D. Kosztin, and K. Schulten. Structural determinants of MscL gating studied by molecular dynamics simulations. Biophys. J. **80**, 2074–2081 (2001) and reprinted with permission from the Theoretical and Computational Biophysics Group at the University of Illinois at Urbana-Champaign.

Fig. 15.11 (a) Adapted from Noguchi, H. and M. Takasu. Fusion pathways of vesicles: a Brownian dynamics simulation. J. Chem. Phys. **115**, 9547-9551 (2001). (b) Courtesy of Dr. Matthias Weiss.

Fig. 16.2 Adapted from Horrobin, D. F. Schizophrenia: the illness that made us human. Medical Hypotheses **50**, 269–288 (1998).

Fig. 17.1 Courtesy of Dr. Katsumi Matsuzaki.

Fig. 17.2 Courtesy of Dr. Roger E. Koeppe II.

Fig. 17.3 Courtesy of Dr. Olaf Sparre Andersen.

Fig. 18.2 Adapted from Zhou, F. and K. Schulten. Molecular dynamics study of the activation of phospholipase A_2 on a membrane surface. Proteins: Structure, Function, and Genetics **25**, 12–27 (1996) and reprinted with permission from the Theoretical and Computational Biophysics Group at the University of Illinois at Urbana-Champaign.

Fig. 18.3 Courtesy of Dr. Thomas Kaasgaard.

Fig. 18.4 Courtesy of Dr. Paavo K. J. Kinnunen and Ms. Tuula Nurminen.

Fig. 19.1 Courtesy of Dr. Russell O. Potts.

Fig. 19.2 Adapted from Norlén, L. Skin barrier formation: the membrane folding model. J. Invest. Dermatol. **117**, 823–829 (2001). Courtesy of Dr. Lars Norlén.

Fig. 19.4 (a) Adapted from http://www.sghms.ac.uk/.../dash/apoptosis/apoptosis.jpg.
(b) Adapted from http://www.ucsf.edu/cvtl/prev/necrosis.html.

Fig. 19.5 Courtesy of Dr. Paavo J. K. Kinnunen.

Fig. 20.1 Adapted from Sackmann, E. and M. Tanaka. Supported membranes on soft polymer cushions: fabrication, characterization, and applications. Trends Biotech. **18**, 58–64 (2000). Courtesy of Dr. Motomu Tanaka.

Fig. 20.2 Adapted from Karlsson, A., R. Karlsson, M. Karlsson, A.-S. Cans, A. Strömberg, F. Ryttsén, and O. Orwar. Networks of nanotubes and containers. Nature **409**, 150–152 (2001). Courtesy of Dr. Owe Orwar.

Fig. 20.3 Adapted from Crommelin, D. J. A. and G. Storm. Magic bullets revisited: from sweet dreams via nightmares to clinical reality. In *Innovations in Drug Delivery. Impact on Pharmacotherapy* (T. Sam and J. Fokkens, eds.). Houten: Stichting Orhanisatie Anselmus Colloquium (1995) pp. 122–133. Courtesy of Dr. Gert Storm.

Fig. 20.5 Courtesy of Dr. David Needham.

Fig. 20.6 Courtesy of LiPlasome Pharma A/S.

Index

AA **XIII**, 174, 175
   and neural synapses 178
   and psychiatric disorder 178
   in blood 156
   in the brain 175
   precursor for eicosanoids 204
Acholeplasma laidlawii 51, 94
actin 212
acyl-transferase 178
acylated polypeptide 141
adaptation 51
adenosine 11
   triphosphate *see* ATP
adipocyte 26
adipose
   subcutaneous tissue 178
   tissue 26, 190
adiposis 149
adrenaline 203
aging 195
AIDS 218
alamethicin 185
alanine 11
alcohol 130
   and anesthesia 182
   effect on protein function 168
   tolerance 184
algae 174
all-*trans* conformation 43
α-helix 40, 139
α-hemolysin 211
α-linolenic acid 173, 175
   food sources for 174
alveoli 115
Alzheimer's disease 195, 226
amino acid 11
   in proteins 38
amorphous solid 92

amphibian 135
amphiphatic 35
amphiphilic 35
amphotericin B 218
analgesic 182
anandamide 215
anchoring
   by hydrocarbon chains 141, 182
   in rafts 127
   of enzymes 166
   of lipids 42
   of peptides 141
   of proteins 88, 117, 139
   strength 42
   via lipid extended conformation 138
Andersen, Olaf Sparre 169
anesthesia 181
   and alcohol 182
   and hydrostatic pressure 182
   clinical concentration 182
   general, mechanism of 182
   lipid mediated 184
   pressure reversal 182
   receptor mediated 183
   theory of 184
anesthetics
   general 182
   local 182, 215
   solubility 182
anisotropic
   material 53
   molecules 92
antagonist 182
anti-cancer drug 207, 222
anti-inflammatory drug 204
anti-oxidant 194, 195
antibiotic 181, 184, 211

antidepressant 182
antifreeze agent 102, 154
antifungal compound 182
antihistamine 182
antimicrobial peptide 185
   barrel-stave mechanism 185
   carpet mechanism 185
antipsychotic 182
antrax 211
apolipoprotein B-100 156
apoptosis 205, 206
   and lipids 205
aquaporin 161
arachidonic acid see AA
archaebacteria 9, 30, 122, 202
Area-Difference-Energy model 67
ascorbic acid 194
aspirin 204
atherosclerosis 149, 195
atomic force microscopy 108, 111, 193
ATP **XIII**, 12
   synthase 212
autism 226
avidin 42, 122, 123

bacteriorhodopsin 89, 142–144, 162
   crystals in membrane 163
Bangham, Sir Alec 216
barrel-stave pore 185
bending
   elasticity 70
   energy 63, 66
   fluctuations 89
   modulus 56, 57, 59, 62, 64, 65, 132, 218
   of membrane 63
β-sheet 40
bicontinuous phase 49, 54, 169
bilayer see lipid bilayer
bile salt 29, 157, 190
bioavailability 214
biological
   fiber 212
   function 19
   function in relation to rafts 127
   solvent 33
biomass 174
biomimetic 209, 210
biopolymer 11, 12

bioprobe force spectrometer 42
biosensor 210
biotin 42, 122, 123
Bloch, Konrad 149
blood
   -brain barrier 181, 214
   capillary 60, 222
   cholesterol transport 156
   clotting 204
   fatty-acid transport 191
   flow regulation 175
   vessel 219
blood cell
   and stem cells 197
   in epidermis 197
   killed by drugs 217
   red 9, 42, 55, 60, 71, 75
   red, cell shape 68, 69
   red, lifetime 60
   red, membrane 57, 68
   red, recycling 197
   red, size 57
   softness 60
   white 9
Bloom, Myer 177
bolalipid 30, 202
Boltzmann's constant (k_B) 5
brain
   -to-body-weight 176
   and intelligence 178
   and mental ill-health 226
   connectivity 178
   development 177
   evolution 175
   fat composition 175
   of lizards 136
   plasticity 177
   size in different species 176
   synapse 178
   under pregnancy 177
Brown, Michael S. 149
butter 23

Ca^{2+}-ATPase 164, 165
Calotes versicolor 135
cancer 215, 218
   and apoptosis 205
   breast 218, 221, 226
   cell 215, 223

colon   195, 226
  therapy   219
  tumor   221, 222
cantilever   140
Cantor, Robert   168
carbon chemistry   10
carboxyl group   24
cardio-vascular disease   226
cardiolipin   30
carpet mechanism   185
catalase   195
catalyst   10, 12, 189
cataract   195
caveolae   127
caveolin   127
Cevc, Gregor   218
cell   2
  animal   13
  archaebacterial   9
  cancer   215, 222
  compartmentalization   13
  corneocyte   198
  cytoskeleton   15
  death   205–207
  differentiation   205
  eubacterial   9
  eukaryotic   9, 15
  growth   166, 212
  hair   212
  locomotion   212
  membrane, model of   38
  molecules of   10
  motility   127, 212
  nucleus   14, 142
  number of   2
  organelle   14
  phantom   212
  plant   13
  proliferation   207
  removal   205
  shrinking in apoptosis   205
  size   9
  surface adhesion   127
  swelling in necrosis   205
  wall   13
cellulose   12
ceramide   28
  and apoptosis   205

and signalling   205
as a second messenger   205
chain lengths   206
in lipid domains   206
in the skin   199
produced by sphingomyelinase   190, 193, 207
propensity for curved structures   206
synthesis   205
cerebral cortex   176
cerebroside   28
chain
  melting   96
  melting, pressure effects   202
  pressure   84
  reaction in peroxidation   195
chemical computation   212
chemotherapy   219–221
Chevreul, Michael E.   149
chimpanzee   176
cholera   211
cholesterol   28, **29**, 31, 164
  and bile salt   190
  and curvature   188
  and diseases   149
  and lipid order   100, 152
  and membrane bending   58
  and membrane softness   61
  and membrane thickness   102
  and protein function   164
  and the liquid-ordered phase   152
  biosynthetic pathway   149, 152, 153
  concentration in different membranes   150
  discovery of   149
  effect on alcohol binding to membranes   131, 184
  effect on membrane thickness   85
  effect on permeability   131
  gradient in secretory pathway   150
  in caveolae   127
  in rafts   125, 126
  in the lung   116
  in the skin   199
  insertion in membranes   131
  mixed with SM and DOPC   122, 126

Index

molecular shape 50
molecular structure 149, 154
phase diagram 102
synthesis 153, 156
transport 127
transport in blood 156
cholesteryl esters 156
choline 27
chromosome 212
cis-double bond 45
cis-fatty acid 45
cisplatin 222
co-enzyme 178
coarsening process 118
coherence length 117, 163
drug effects 163
temperature effects 163
collective
diffusion 88
phenomena 91
colloid 53
colloidal
force 58
particle 58
combination therapy 222
complex systems and complexity 17
compressibility
area 56, 59, 63, 132
bulk 53
computer
simulation 78, 81
technology 212
condensed phase 53
conformational degrees of freedom 93
cooperative
behavior 33, 118
phenomena 91
COPI vesicles 172
COPII vesicles 172
Cornell, Bruce 211
corneocyte 197, 198
coronary heart disease 149, 226
cosmetics 209
covalent bond 5
cow 175, 176
Crawford, Michael 176
critical
mixing 62

phenomena 94
critical point
in lipid bilayer 132
in lipid monolayer 108, 111, 112
in lipid-cholesterol mixtures 102
in lipids 100
of water 94
crystal
lipid bilayer 96
liquid 93
protein 49, 143, 162, 169
cubic
phase 49, 54, 215
phase and protein crystallization 169
phase, viscosity of 215
structure 49, 65
structure in the skin 198
Curie temperature 94
curvature 63, 113
and cholesterol 188
and coupling to in-plane degrees of freedom 105
and membrane function 52
Gaussian 64
in membrane fusion 171, 172
induced by ceramide 206
intrinsic 66
local 64, 66, 171
mean 56, 64
of interface 56
of membrane 50
radius of 63
spontaneous 47, 51, 66, 67, 84
stress and channel function 188
stress field 47, 51, 84
stress field, and enzyme attack 192
stress in extremophiles 201
stress release near proteins 167
stress, dependence on head group size 167
cyanobacteria 150
cycline 215
cystic fibrosis 17, 219
cytochrome c 127
binding to membranes 139, 167
cytosis 127
cytoskeleton 15, 55

of red blood cell   61
anchoring of proteins   88
elastic deformation   68
network   212

DAG   **XIII**, 25, 26
in signaling   204
Danielli-Davson model   75
DAPC   **XIII**
bilayer lateral structure   119
bilayer mixed with DLPC   122
bilayer mixed with DPPC   123
Darwinian selection   150
data-driven science   3
DCPC   **XIII**
de-wetting   41
deformability   53
degree of freedom   92, 103, 105
δ-lysin   185
dendrite   110
deoxyribonucleic acid   *see* DNA
dermal
barrier   181, 218
route   200
dermis   197, 198, 200, 218
desaturation enzyme   173
design
bottom-up   225, 226
principles for natural materials   209
desmopressin   182
detergent   35, 54, 124
-resistent membrane fraction   124
and protein crystallization   169
DGDG   **XIII**
DHA   **XIII**, 25, 174, 175
brain function   177
curvature stress   168
in the brain   175
neural synapses   178
psychiatric disorder   178
rhodopsin function   168
di-acylglycerol   *see* DAG
di-arachioyl PC   *see* DAPC
di-biphytanyl-diglycerol-tetraether   30
di-decanoyl PC   *see* DCPC
di-laureoyl PC   *see* DLPC
di-myristoyl PC   *see* DMPC
di-myristoyl PE   *see* DMPE
di-myristoyl PG   *see* DMPG

di-oleoyl PC   *see* DOPC
di-palmitoyl PC   *see* DPPC
di-palmitoyl PE   *see* DPPE
di-stearoyl PC   *see* DSPC
di-stearoyl PE   *see* DSPE
dielectric constant   129
diet
and fatty acids   173
carbohydrates   226
fish-rich   226
Icelandic   226
Japanese   226
Mediterranean   226
dietary fats   173
diffusion
and cholesterol   152
and disorder   103
and lipid domains   124
and permeation   129
in lipid annulus   162
lateral   87
of lipids   87, 88, 96
of nicotine   200
of proteins   87, 88, 124, 163
of water   42, 68, 168
rotational   87
digalactosyl diglyceride   *see* DGDG
digestion   191
dipole-dipole interaction   33
discocyte   70
dismutase   195
disorder   16, 18
dispersion   36
DLPC   **XIII**
bilayer mixed with DAPC   122
bilayer mixed with DPPC   122
bilayer mixed with DSPC   100, 122
DMPC   **XIII**, 25, 60, 107
area compressibility modulus   132
bending modulus   132
bilayer lateral structure   119
bilayer mixed with DMPE   122
bilayer mixed with DSPC   119, 122, 123
bilayer mixed with other lipids   99
bilayer permeability   130
bilayer thickness   86
enzyme binding   141

ethanol binding   130
lipid monolayer structure   111
monolayer isotherm   108
monolayer lateral structure   112
monolayer mixed with DPPC   113
multi-lamellar repeat distance   133
phase transition   120
phospholipase action   134
protein binding   140
softness   60
specific heat   100
DMPE **XIII**
  bilayer mixed with DMPC   122
DMPG **XIII**
  enzyme binding   141
DNA **XIII**, 11, 12, 195, 212
  damage by drugs   219
  encapsulation   218
  gene therapy   218
docosahexaenoic acid   *see* DHA
docosapentaenoic acid   *see* DPA
dolphin   176
domains   *see* lipid domains or membrane domains
DOPC **XIII**
  in rafts   125, 126
  mixed with SM and cholesterol   122, 126
  trans-bilayer profile   82
double
  bond   24, 31
  bond alteration by enzymes   173
  bond migration   195
  helix   12
doxorubicin   218–220, 222
DPA **XIII**, 175
  in the brain   175
DPPC **XIII**
  and phospholipase action   134, 193
  area compressibility modulus   132
  bilayer configuration   97
  bilayer lateral structure   119
  bilayer mixed with DAPC   123
  bilayer mixed with DLPC   122
  bilayer mixed with DPPE   122
  bilayer permeability   131
  bilayer thickness   86
  bilayer with cholesterol   101
  binding of acylated polypeptide   141
  in the lung   114
  monolayer lateral structure   112
  monolayer mixed with DMPC   113
  monolayer on gas bubble   107
  phase transition   95, 120
  phase transition in the lung   116
  ripple phase   98
  specific heat   95
  trans-bilayer profile   82
  transition entropy   96
DPPE **XIII**
  bilayer mixed with DPPC   122
drug
  action sites   181
  anti-cancer   218
  carrier   216
  degradation   214
  delivery   209
  delivery by liposomes   214
  depot   214, 215
  dermal application   200
  design   209
  encapsulation   214
  enhancer   200, 223
  formulation   181, 214
  free   219
  hydrophilic   214
  hydrophobic   214
  perfect   216
  potency of   182
  release   214, 221
  release by enzymes   221
  solubility   181
  target   214
  targetting   216, 217
DSPC **XIII**, 44
  and phospholipase action   134
  area compressibility modulus   132
  bilayer lateral structure   119
  bilayer mixed with DLPC   100, 122
  bilayer mixed with DMPC   119, 122, 123
  bilayer mixed with other lipids   99
  phase transition   120
  specific heat   100
DSPE **XIII**, 44
dynamic heterogeneity   118

dyslexia   178

echinocyte   70, 71
egg yolk and white   35, 53, 174
eicosanoid   175, 204, 226
eicosapentaenoic acid   *see* EPA
elastic
   spring   56
   stress in membranes   170
electric field-induced pore formation
   200
electrical-optical transducer   211
electron transport   139
electronics   209
electrophysiology   170
electrostatic interaction   110
elongation enzyme   173
emergent phenomena   17
emulgator   36
emulsion   36, 53
   and enzyme attack   190
   in the stomach   190
encapsulation   10, 15
endocytosis   171
endoplasmic reticulum   *see* ER
endothelia   127
energy
   bending   57
   free energy   58
   free energy of transfer   42
   Gibbs excess free energy   55
   mechanical   58
   turnover in body   164
enthalpy   58
   of lipid phase transition   96
entropic
   force   58, 59
   repulsion   131, 217
entropy   18, 33, 49, 53
   Boltzmann's formula   96
   configurational   58
   of lipid phase transition   96
   of soft matter   58
enzyme
   action induced by micro-manip-
      ulation   194
   activation by curvature stress   166
   activation by lipids   134, 166
   activation by products   192
   and phase transitions   133, 134
   as catalyst   189
   attack at interface   190
   binding to membrane   140
   digestive   189
   drug inactivation   181
   lipase   189
   lipolytic   26
   phospholipases   189
   regulation of synapse formation   178
   sphingomyelinase   189
   turnover rate   192
   use in drug delivery   221
EPA   **XIII**, 174
   for treating diseases   215
epidermis   197, 198
ER   **XIII**, 14, 164, 204
   cholesterol content   150
   exit sites   171, 172
   membrane   31, 50, 58
ergosterol   29, 102, 150
Erlich, Paul   216
erythrocyte   *see* blood cell, red
Escherichia coli   159, 170
Escherichia coli   9, 13, 14, 50, 52
essential fatty acid   173
   digestion   191
ester bond   24
ethanol   130, 182
   binding to membranes   130, 183
   shift in lateral pressure profile   183
ethanolamine   27
ether
   bond   30, 202
   lipid   221
eubacteria   9
eukaryote   9
Evans, Evan   42, 212
evolution   2, 16, 31
   Homo sapiens   176, 179
   and chemistry   176
   and materials design   19, 209
   and oxygen pressure   151
   chemical   152
   common ancestor   9
   conservation of protein sequences
      142
   importance of fatty acids   176

importance of water   41
of proteins   165, 178
of sterols   149
of the human brain   175
exocytic pathway   165
exocytosis   166, 171, 198
extremophil   201

fat   11, **23**
dietary   157
globule   190
in milk   190, 191
solubility in water   23, 190
storage   26
fatty acid   XIII, 11, **24**, 25
ω-3   173
ω-6   173
cis-   45
n-3   173
n-6   173
trans-   45
branched   30
carboxyl group   24
chain anchoring to membranes   141
chain flipping   166
classification   24
degree of saturation   24
desaturation   175
elongation   175
essential   173
free   24
from the diet   173
in drug delivery   222
in fish   176, 177
in the skin   199
modified by acyl-transferase   178
polyunsaturated   24, 173, 174, 202
solubility   24
substitution   173
super-unsaturated   24, 174
synthesis   24
transformations   173
transport by lipoproteins   178
transport by proteins   24
transport in blood   191
fatty-acid
chain length   24, 31
fetus   177
fiber   12

film balance   105, 107
fish
and fatty acids   177
as food source   175
egg   177
embryo   177
fat   175
lipid composition   176
oil   23
smelly   194
flagella   212
flip-flop process   87
flippase   87
flotation   209
fluctuations
and bilayer heterogeneity   132
and domain size   163
at a phase transition   94
at lipid phase transition   96, 132
dependence on lipid type   132
destabilization of lamellar structure   65
in bilayer thickness   132
in composition   61
in density   61, 118, 134
in lipid monolayer   108
in vesicle shape   68
of liposome   60
of membranes   58
of surface   59
fluid-mosaic model   74, 75
fluidity   75
fluorescence
correlation spectroscopy   87
energy transfer   118, 120
labelling   87
microscopy   121, 122
microscopy, length scale   110
fluorescent
acceptor   118
donor   118
molecules   110, 118
foam   209
folding
of lung linen   115
of membrane   76, 198
of membrane proteins   144
of proteins   38, 142

reverse folding of proteins   145
  unfolding of proteins   145
food
  contamination   194
  effect on membrane composition   51
  essential fatty acids   173
force
  between membranes   131
  between soft interfaces   58
  colloidal   49, 58
  covalent bond   5
  electrostatic   38
  entropic   58, 132
  hydrogen bond   5
  hydrophobic   139
  mechanical   58
  physical   5, 19, 138
  thermodynamic   58
  transducer   42
  van der Waals   5
fractal   110, 114
frog skin   185
fungus   9, 29, 194
fusion
  of membranes   170

G-protein   177
gall
  bladder   190
  stone   149
ganglioside   28
gas   92
  as anesthetic   182
  bubble   107
  ideal   58
gastrointestinal epithelium   214
Gaub, Hermann   143
Gauss-Bonnet theorem   65
Gaussian curvature   64
gene   11
  and evolution   152
  and psychiatric disorder   178
  defect   178
  delivery   218
  expression   3, 204, 215
  modification   17
  number of   17
  therapy   17, 218
genetic information   11, 17

genetically determined disease   17
genocentric   3
genome   11, 16
  coding for integral membrane
    proteins   137
  coding for water soluble proteins
    169
  human   12
  human genome project   16, 226
genomics   3
  biophysical   17
genotype   18
genus number   64, 65
gestation period   177
glass   92
  bead   42
glucose   11
glue   53
glutathione peroxidase   195
glycero-phospholipid   26, 27
glycerol   24
  backbone   24, 183, 189
  transport in proteins   159
glycocalyx   15, 76
glycolipid   26, 27, 31
glycosphingolipid   184
  in rafts   125
glycosylation   27
Goldstein, Joseph L.   149
Golgi   165
  apparatus   14
  cholesterol content   150
  membrane   14, 50, 58, 198
  protein transport   164
  vesicle transport   171
Gorter and Grendel model   75
grain boundary   107
gramicidin A   169
  dimer formation   169
  in nano-electronics   211
  ion conductance   186
growth factor   166, 204

H_{II} phase   49, 51, 94
  and gramicidin A activity   170
  and protein function   166
  in signalling   204
  induced by ceramide   206
  promotion   50

H_I phase 49, 94
hairpin conformation 141
hard matter 53, 209
HDL **XIII**, 157
heart attack 157, 226
Helfrich, Wolfgang 58, 63
high-density lipoprotein see HDL
homeostasis 52, 94, 136, 201, 202
Hooke's law 56
hopanoids 156
hormone 166
  and signalling 203
  melatonin 195
  peptide 182
  precursors for 175
  transport 156
Horrobin, David 178
hydrocarbon
  chain 23
  surface 41
hydrocarbon chain
  in water 35
  saturated 23
  unsaturated 23
hydrogen bond 33
  network 33, 34, 40
hydrogen peroxide 194
hydrolysis 24
hydrolysis products 193
  as drugs 221
hydrolytic cleavage 191
hydrophilic **24**, 34
  surface 41
hydrophobic 23
  crevice 166
  effect 34, 40, 42, 166
  matching 159, 160
  matching and protein function 163
  sites for drug action 181
  surface 41
  thickness 85
  thickness profile 161
hypercholesterolemia 157
hypertension 226
hyperthermia 219
hypothalamus 136
hypothesis-driven science 3, 18

ice

amorphous 41
density 40
freezing-point depression 91
melting 91
ideal gas 59
imaging
  of lipid bilayers 121
  techniques 121
immune system 175, 217
immunodepression 184
implants 210
inflammatory response 204
information science 3
inositol 27
  in signalling 204
insulin 200, 203
integral membrane protein 138, 160
interface
  bending 56, 63
  between oil and water 35, 36
  between two liquids 54
  between water and air 37, 105
  compression 56
  enzymatic attack 190
  fluid 56, 66
  forces between 58
  hydrophilic-hydrophobic 82
  mathematical 64
  mechanical properties 56
  softnesss 58
  stiffness 55
  structure 41
  target for drugs 181
  tethered 55
  topology 64
  tryptophans at membranes 141
  with internal structure 66
interfacial tension see tension, interfacial, 35
interfacially active molecule 54, 55, 183
interfacing hard and soft matter 209
intestinal barrier 181, 191
intestine 190
intracellular communication 212
inverted hexagonal phase see also H_{II} phase, 49, 215
ion

conductance of gramicidin A   186
conductance of membranes   170
  permeability   131
  pump   164
  transport   164
Ipsen, John Hjort   101
isopranyl   127, 202
isotherm of monolayer   108
Israelachvili, Jacob   75

Jørgensen, Kent   221

keratin   198
keratinocyte   197
ketchup   53
kinesin   212
kingdoms of life   9
Kinnunen, Paavo   167, 206

lag phase of enzyme action   134
lamellar
  body   115, 198
  phase   54
  repeat distance   86, 133
  structure   38, 65
lamin   142
land mammals   177
Langmuir
  -Blodgett film   107, 109, 121
  film   105–107
  trough   106
lanosterol   152
  molecular structure   154
Laplace pressure   113
lard   23
lateral
  diffusion   87, 96
  pressure profile   83, 84, 138, 165
  pressure profile and alcohol   183
  pressure profile and protein function   168
latex bead   192
laughing gas   182
LDL   **XIII**, 156
  receptor   156
lecithin   35
leukotriene   175, 204
lidocaine   215
life

evolution of   151
  extraterrestrial   33
  importance of water   33
  kingdoms of   9
  molecules of   10, 11
  origin of   10, 24
  under extreme conditions   200
light-sensitive protein   167
Lindblom Göran   4, 51
linear molecule   11
linker molecule   140
linoleic acid   173, 175
  food sources for   174
lipase   189
  digestive   191
  in skin   198
lipid   XIII
  aggregate   37, 47
  all-*trans* conformation   43
  anchoring strength   42
  and anti-oxidants   194
  and apoptosis   205
  and diet   226
  and disease   177
  and fats   1
  and genes   1, 3
  annnulus and drugs   162
  annulus around protein   160
  archibacterial   122
  as drug   215
  as energy source   193
  as messenger   203
  asymmetry   31, 205
  backbone of   24
  barriers for drug action   181
  bilayer   11–14, 37, 38
  bilayer as insulator   210
  bilayer configuration   97, 119
  bilayer curvature   66
  bilayer defects   130
  bilayer fusion   170
  bilayer instability   166
  bilayer lateral structure   122
  bilayer on support   123, 193, 211
  bilayer permeability   129
  bilayer phase transition   94
  bilayer softening   131, 133
  bilayer stability   78

bilayer stress   84
bilayer thickness   13, 45, 63, 85, 86, 97, 138
bilayer thickness, importance for lipid-protein interactions   159
bilayer, closed   67
bilayer, multi-lamellar   37
bilayer, softness of   43
bipolar (bolalipid)   30
chain flexibility   43
chain melting   96
changing shape   48, 49
charged   26, 31, 127, 139, 141
conformation   43, 45
conformation in raft   125
conformational change   87
cross-sectional area   45, 46, 97, 108
crystalline domains in skin   200
cubic structure   49
curvature   43, 46, 47
curvature stress   48, 167
de-mixing by proteins   160
dehydration   49
di-ether   30
diffusion and domains   124
diversity   16, 19
domain size   118
domains   89
domains and enzyme activity   134, 192
domains and protein binding   140
domains around proteins   163
domains in bilayers   117, 118, 120, 121, 123, 133
domains in monolayers   111
domains in the lung   116
dynamics   81, 86, 118
effective shape   46
enzyme activity on mixtures   135
enzyme degradation   189
ether   221
extended conformation   138, 167
fats   23
fatty-acid chain length   43
flip-flop   87
for micro-encapsulation   214
going rancid   193
head group   26

hydrophilic head   26
hydrophobic tail   26
in drug delivery   214
in evolution   152
in lung function   112
in nanotechnology   209
in the brain   175
in the skin   197
intestinal degradation   191
lateral diffusion   87, 88
low-technological applications   209
membrane composition   31
mixtures   98, 120, 122
molecular degrees of freedom   96, 103
molecular shape   45, 46, 66, 84
molecular volume   46
molecule   24
monolayer   37, 46
motion, time scales of   86
neutral   26
non-lamellar aggregate   46
non-lamellar structure   49
non-lamellar structure and function   169
nonpolar   24
oxidation   193, 194
packing parameter   46, 47, 49
PC family   44
peroxidation   194
perturbation by proteins   159
phase diagram   99
phase equilibria   99
phase transition   119, 129, 132, 220
phase transition and enzyme activity   134
phase transition temperature   96
phases   97
physical properties of   4
polar   26, 27
polar head group   XIII
poly-isoprenoic   30
polymorphism   49, 215
pressure effect   49
protrusion out of the bilayer   87
raft   122, 124, 125, 207
regulation of composition   135
remodelling   189

removal of head group 48
rotational dynamics 87
signalling 26, 204
size of molecule 43, 44
sorting 127
sorting at protein surface 160
spontaneous breakdown 194
spontaneous curvature 47
storage 26
supported bilayer 123
synthesis 24, 31, 52
target for drugs 181
temperature effect 43, 48, 85
tetraeter (bolalipid) 30
transport across membranes 189
tube 49
vesicle 54
volume change 49, 97, 202
zwitter-ionic 26
lipid bilayer membrane *see* lipid bilayer
lipid II 186
lipid-protein interactions 142, 159
lipidomics **1**, 3, 17
LiPlasome 222
lipolysis 26, 189
lipopolymer 58, 217
lipoprotein 24
  and brain formation 178
  transport of fatty acid 191
liposome 37–39
  and hyperthermia 220
  circulation time in blood 218
  discovery 216
  first-generation in drug delivery 217
  in blood capillary 216
  in drug delivery 214, 216, 220
  leakiness 219, 221
  micro-manipulation of 213
  multi-lamellar 37, 39
  phase transition 220
  problems in drug delivery 219
  second-generation in drug delivery 217
  size 63
  transfersome 218
  uni-lamellar 39
  with ceramide 207

liquid 53, 75, 92
  structured 53
liquid crystal 53, 93
  nematic 93
  smectic 93
  technology 93
liquid-disordered phase 95–97, 103, 118
liquid-ordered phase 101, 103, 118, 125, 154
  induced by cholesterol 102
liver 157, 177
living matter 209
lizard 135
low-density lipoprotein *see* LDL
lubrication 209
lung 112
  alveoli 112
  collapse 114
  mechanics 113
  protein SP-B 115
  protein SP-C 115
  pulmonary surfactant 114
  surfactant proteins 114
lymphatic system 218
lyso-ether lipid 222
lysolipid 25, **26**, 48, 191
  and fusion 172
lysosome 14, 156, 172
lysozyme 40

Müller, Daniel 143
macromolecular
  assembly 11
  branched network 12
macromolecule 11, 12
macrophage 217
macroscopic world 4
magainin 2 185
magic bullet 216, 221
magnetic resonance techniques 81
main phase transition 85, 95
  temperature 86
manic-depression 178, 226
Margulis, Lynn 151
marine fouling 213
Mars 24
materials
  design 19, 209

natural vs. man-made   60
mayonnaise   35
mean curvature bending modulus   64
mechanical
 force   58
 modulus   56
 strain   170
mechano-sensitive channel   170
melatonin   195
melittin   185
membrane   2
 -spanning protein   142
 adsorbed   61
 alcohol binding   183
 anchoring of proteins   138
 and non-lamellar structure   50
 animal plasma   31
 archaebacterial   30
 bending rigidity   58
 biological   13, 15, 18
 blebbing   205
 channel   164, 169
 chloroplast   30
 compartment   128
 composition   14, 51
 curvature stress   167
 defects and permeability   130
 differential area   67
 domain size   121
 domains   122, 123
 domains and diffusion   88
 domains and protein binding   140
 domains around proteins   163
 dynamics   88
 effect of cholesterol   85
 elastic stress monitoring   170
 fluidity   75, 76, 86, 97, 102, 103, 201
 folding   76, 199
 function related to rafts   125
 fusion   170
 Golgi   31
 heterogeneity   77, 83, 117, 133
 history of models   74, 76
 hydrophobic core   82
 hydrophobic thickness   203
 in nanotechnology   209
 interaction with proteins   162
 internal   14, 58

invagination   172
lateral diffusion   152
lateral structure   15, 117, 119
lipid mixtures   98
lipid transport   189
mechanical stability   154
mitochondrial   30, 31
model of   73, 88, 89
neural   175
number of lipids   13
of extremophill   200, 201
of pulmonary surfactants   116
on solid support   123, 193, 211
organelle   31
organizational principles   81
partitioning of solutes   131
percolation of lipid domains   128
permeability   129
physical properties optimized   152, 177
plasma   13, 14
pore   171
pore formation   200
proteins   138
raft   122, 124, 125
raft structure   125
red blood cell   68
retinal rod outer segment   168, 176
rupture   59
shape   67
shear resistance   63
softness   59
stalk   171
stalk formation   171
surface undulations   58
target for drug   181
thickness   24, 81, 84, 85
thickness and protein sorting   165
thickness, effect of cholesterol   102
thickness, importance for lipid-
 protein interactions   159
thinning by proteins   159
transport   129
transport by vesicles   172
tryptophans at interface   141
unbinding   61
undulations   132
with bolalipids   203

mental
  disease 226
  ill-health 226
meso-phase 93
messenger 12
  and apoptosis 205
  lipid 203
metabolic syndrome 226
metabolomics 3
metallization 213
methane 202
Methanococcus jannischiiwas 9
metronidazole 215
Meyer-Overton relation 182
Miao, Ling 67
micellar phase 215
micelle 37
  in milk 190
micro-electronics 109
micro-emulsion 54, 59
micro-manipulation 212
micro-pipette aspiration
  of monolayer 107
  of bilayer 60
micrometer (μm) 5
microscopic world 4
microtubules 212
milk
  fat 190, 191
  mother's 177, 191
  sour 9
minimal surface 49, 64, 65
mitochondria 14, 139
  cholesterol content 150
  ROS production 195
mixture
  of lipid and water 37
  of lipids 98, 119
  of oil and water 36
model
  of biological membrane 73
  relevant variables 74
model system 73
molecular
  -probe experiments 81
  force transducer 170
  fossil 152
  long axis 92

motor 212
pore 159
smoothness 152
structure 1, 15, 16, 18
Molecular Dynamics simulation 78
molecule
  prolate 93
  spherical 92
mono-acylglycerol 191
monolayer 37
  area 106
  enzyme attack 191
  fluctuations 108, 112
  fluorescence microscopy 110
  isotherm 107, 108
  lateral pressure 106
  lateral pressure vs. hydrostatic
    pressure 107
  lateral structure 108, 110
  lipid 31, 37, 66, 105
  lipid domains 110
  lung surfactant 115
  of membrane 71, 81
  on gas bubble 106
  on solid support 109, 111
  phase transition 105, 107
  phases 108, 111
  spontaneous curvature 47
  thickness 113
monomer 11
monomolecular film 105
monosaccharide 12
Monte Carlo simulation 78
mouse 176
multi-drug resistance 184
multi-lamellar lipid bilayers 86, 133
multidisciplinary science 18
muscle
  cell 164
  contraction 204, 212
mycoplasma 51
myoglobin 12
myosin 212
myristic acid 25

n-3 fatty acid 173
n-6 fatty acid 173
Na^+, K^+-ATPase 165
nano

-actuator   212
-biotechnology   210
-electronics   109, 212
-machine   170
-meter (nm)   5
-meter-scale conduits   212, 213
-scale   5
-scale materials design   210
-science   20, 225
-technology   20, 209
-tubes, carbon   213
-tubes, lipid   212
-tubular network   213
NASA   33
necrosis   205, 206
Needham, David   220
nerve
  cell   130
  signal   172
network
  tethered   55
neural
  growth failure   178
  membrane   175
  synapse   178
  system
    and alcohol   182
  system in fish   177
  tissue, lipid composition   176
neuron   178
neurotransmitter   166, 204
neutron scattering   81
nicotine   200
nisin   185
Nobel Prize   149, 156
non-communicable disease   226
non-lamellar phase   47, 49, 50
  and drug delivery   223
  and drug release   215
  and protein function   166
  in membrane fusion   171
  induced by enzyme action   192
Norlén, Lars   198
nucelotide   11
nuclear envelope   14
nucleation   169
nucleic acid   11
nut   194

nutritional value   194

obesity   226
oil   23
  burning   193
  fish   23
  liquid   34
  olive   23, 226
oleic acid   11, 25, 51
ω-3 fatty acid   173
ω-6 fatty acid   173
optimization via evolution   152
order   15
organocentric   3
organogenesis   205
orientational degrees of freedom   93
osmotic pressure   58, 68
Ourisson, Guy   155
outer membrane protein A (OmpA)   144
oxidation
  of lipids   193
  of unsaturated lipids   194
  toxic products   194
oxidative stress   195
oxygen
  and lipid degradation   194
  in the atmosphere   150
  partial pressure   151

paint   53
palmitic acid   51
palmitoyl-oleoyl PC   *see* POPC
palmitoyl-oleoyl PE   *see* POPE
palmitylation   182
pancreas   191
Parkinson's disease   226
partition coefficient   182
pattern formation   110, 119
PC   **XIII**, 27, 31, 165
  and fusion   172
PE   **XIII**, 27, 50
  activation of proteins   167
  and fusion   172
  and rhodopsin function   168
penicillin   184
peptide   11
  acylated   141
  antimicrobial   185

peptidoglycan layer 14
percolation 128
perinatal condition 226
periodontitis 215
peripheral membrane protein 139, 166
permeability
  barrier 129
  barrier of the skin 200
  of ions across bilayers 130
peroxidation 194
persistence length 55, 57
PG **XIII**, 27
  in the lung 114
pharmacokinetics 214
phase diagram 99
  vesicle shapes 68
phase separation 98, 100, 123
  dynamics of 119
phase transition 91
  and bilayer permeability 130
  and biological function 129
  and enzyme action 133
  and solute binding to bilayers 130
  and undulation forces 133
  as a thermometer 135
  continuous 94
  discontinuous 93
  enthalpy 96
  entropy 96
  first order 93
  in drug delivery 220
  in lipid monolayer 105
  in lizards 135
  in two dimensions 105
  in vesicle shape 68
  lamellar–non-lamellar 50
  lipid bilayer 94
  lyotropic 94
  red blood cell shapes 71
  second order 94
  temperature 94, 96
  thermotropic 91, 93
  vesicle budding 70
phases of matter 53
phenotype 18
phosphatidic acid 25, 26
phosphatidylcholine *see* PC
phosphatidylethanolamine *see* PE

phosphatidylglycerol *see* PG
phosphatidylinositol *see* PI
phosphatidylserine *see* PS
phospholipase 27, 189
  activity dependence on lipid substrate 134, 192
  in drug release 221
  in signalling 204
  phospholipa+-se A_2 204
  phospholipase A_1 189
  phospholipase A_2 48, 133, 134, 140, 141, 178, 189, 191, 221, 222
  phospholipase B 189
  phospholipase C 48, 178, 189
  phospholipase D 189
  secretory 134
  upregulation 221
photo-induced cross-linking 213
photosynthesis 150
photosynthetic reaction center protein 143
phylogenetic tree 10
physical force 5
physical principles 19
physics 19
phytanyl 202
phytoplankton 174
PI **XIII**, 27, 31
  in signalling 204
pico-Newton (pN) 5
piezophile 201
placenta 177
plant
  green 174
  seeds 174
plasmid 218
plastic 60
platelet 9
plumber's nightmare 54
polar
  group 26
  lipid 27
poly-isoprene 202
poly-peptide
  membrane binding 141
polyethylene 60
polymer 53, 55
  coat 217, 222

confinement   59
cushion   58, 210
flexibility   55
phase   54
surface coating   59
water soluble   217
polymerization   213
polymorphism   49
polynucleotide   11
polypeptide   11
   membrane spanning   163
polysaccharide   11, 12
polyunsaturated
   fatty acid   173
   lipid   24
Pond, Caroline   225
POPC   **XIII**
   bilayer mixed with POPE   88
   bilayer thickness   85
   bilayer with aquaporin   161
POPE   **XIII**
   bilayer mixed with POPC   88
   bilayer with aquaporin   161
pore formation   200, 211
   in membrane fusion   171
porin   40
post-genomic era   16
post-translational modification   17
pre-transition   95, 98
pregnancy   177
pressure
   and anesthesia   182
   and life at deep sea   200
   and piezophiles   201
   effect on phase transition   132
   hydrostatic   49, 201
   hydrostatic vs. lateral   107
   in lung   114
   in micro-pipette aspiration   60, 106
   Laplace   113
   lateral   168
   lateral pressure profile   83
   of Langmuir film   105
   of oxygen in the atmosphere   151
   osmotic   58, 68
   reversal phenomenon   182
   two-dimensional   105
prodrug   216, 221

proenhancer   223
programmed cell death   205
prokaryote   9
   membrane   31
prostaglandin   175, 204
protein   11
   α-helical   142
   -protein attraction in membranes   162
   -protein interaction mediated by lipids   160
   acylation   127
   amphiphilic   143
   anchored to cytoskeleton   117
   anchoring at membranes   138, 139
   as molecular motor   212
   at membranes   137, 139
   binding to lipid domains   127
   conformational change   78, 168, 169
   crystallization   169
   denaturation   38, 169, 194
   electrostatic binding   138
   folding   38, 142, 144, 145
   function   17, 137
   function and cholesterol   165
   function and membrane thickness   165
   function and the lateral pressure profile   165
   function by curvature stress   168
   helix bundle   170
   hydrophobic crevice   166
   in association with rafts   127
   in solid membranes   163
   insertion in membrane   144
   kinase C   127, 166, 204
   lateral diffusion   88, 124, 127
   lateral organization in membranes   163
   light-sensitive   167, 177
   membrane association in models   75
   membrane spanning   142
   membrane-spanning domain   138, 142
   peripheral to membranes   139
   positively charged   140
   rotational motion   88
   secretion   145, 164

shape change   169
signalling   207
solubility   137
solvation by ions   166
sorting   14, 150, 164
synthesis   164
three-dimensional structure   142, 143, 169
transport   14, 145
two-dimensional crystals   143, 144, 162
water soluble   169
protein-lipid interactions   *see* lipid-protein interactions
proteolipid complex   172
proteomics   3
proton
  pump   143, 212
  transport   143
PS   **XIII**, 27, 31, 166
psychiatric disorder   178
psycrophil   201
pulmonary surfactant   114

radius of curvature   56, 64
  of membrane   63
raft   124, 125
  and lipid sorting   127
  and membrane function   125
  around integral protein   160
  caveolae   127
  in lipid mixtures   126
  in liquid-ordered phase   125
  in relation to function   127
  in signalling   207
  in the secretory pathway   165
  lipid extract   122
  mixture   126
  protein association   127
  schematic illustration   125
  sizes   126
rancid   193
randomness   77
rat   176
reactive oxygen species   *see* ROS
receptor
  avidin   42
  calcium-activated   172
  co-localization with ligands   128

drug binding   137
for drugs   181
in general anesthesia   183
in imaging   121
in sensor   210, 211
in signal transduction   203
LDL   156
lipid   3
lipoprotein-binding   149
membrane-bound   214
membrane-spanning   142
surface-cell   212
recognition system   12
reptile   176
respiration   151
respiratory distress syndrome   114
retina   167
  lipid composition   176
rhinoceros   176
Rhodopseudomonas viridis   143
rhodopsin   167, 177
ribonucleic acid   *see* RNA
ribosome   164
Rilfors, Leif   4
ripple phase   95, 98
  periodicity   98
RNA   **XIII**, 10, 11
Robertson's unit membrane model   75
robustness   18
ROS   **XIII**, 194
rubber   55

Sackmann, Erich   76
saddle-splay modulus   64
saliva   191
salmon   175
saturated lipid   23
saturation of hydrocarbon chains   23
schizophrenia   177, 226
  and fatty acids   178
schizotypy   178
second messenger   204
secretory
  pathway   150, 164
  pathway and cholesterol   165
  pathway and rafts   165
  phospholipase   134
self-assembly   2
  driving force for   33, 34

in drug delivery   220
   of lipids   37
   of soft materials   209
self-healing   38, 209, 210
sensor   210, 211
serine   27
sex hormone   29
shear resistance   55
shell fish   175
sieving mechanism   165
signal
   molecules   203
   pathway   203
   transduction   127, 166, 203
signalling   203
   cascade   203
   in relation to rafts   127
   pathway   205
   process   77
silly putty   53
Singer-Nicolson model   75
single-channel activity   170
single-molecule
   enzymology   192
   force spectroscopy   42, 140, 144
   tracking   87, 124
sitosterol   29, 31, 150
skin
   as a barrier   197
   ceramide   190
   cholesterol   199
   lipids   197
   model, brick and mortar   197
   model, domain mosaic   200
   nasal   200
   permeation by transfersomes   200
   structure   198
SM   **XIII**, 26, 28, 31
   and membrane thickness   165
   in rafts   125, 126
   mixed with DOPC and cholesterol   122, 126
   precursor for ceramide   205
   regulation   205
SNARE-protein   172
soap   35, 54
   bubbles   114
soft
   interface   19, 53, 54, 57, 132
   liposome   57, 218
   materials   209
   matter   2, 19, 53, 54
   matter in drug delivery   220
softness
   and lipid chain type   61
   of materials   53
solid   53, 92
solid-ordered phase   95–97, 118
solid-solid phase separation   100
SOPC   **XIII**, 44, 45
sorting
   of lipids by proteins   160
specific heat   95
   and lipid transitions   100
   of lizard membranes   136
   peak in   95
spectrin   55, 68
sphingolipid   28
   in rafts   125
sphingomyelin   *see* SM
sphingomyelinase   27, 189, 194, 205
   in the skin   199
sphingosine   26
spinal cord   197
sponge phase   53, 54, 65
   in the lung   115
spontaneous
   curvature   51
   process   37
squalene   152
squirrel   176
starfish shape   70
states of matter   92
statin   157
stealth liposome   58, 217, 221
stearic acid   25
stearoyl-oleoyl PC   *see* SOPC
stem cell   197
steroid skeleton   28, 100, 152
sterol   29
   evolution   149
   methyl groups   152
stomach   190
stomatocyte   70, 71
storage lipid   26
stratum corneum   197, 198

stress *see* curvature
stress profile   84
stroke   226
structure-function relationship   16, 18, 162
sugar   11, 12
  burning   194
  on lipids   26
  transporters   164
sulphur   202
  bridge   38
super-unsaturated
  fatty acid   173
  lipid   24
surface
  coating   209
  fluctuations   59
  functionalization   212
  minimal   64
  undulations   58
surfactant   36
sustainability   225
symbiosis theory   151
synapse   178
syrup   53

tension
  interfacial   35, **55**, 84, 105, 114
  lowering of interfacial   36
  surface   35, 212
  surface, of water   105
  vesicle   50
testosterone   29
tetra-ether   202
therapeutic effect   220
thermal
  energy ($k_B T$)   5
  renormalization   132
thermophile   201
Thermoplasma acidophilum   203
thromboxane   204
tissue remodelling   205
toothpaste   53
topology
  invariant   64
  of membrane   58, 64, 66
toroidal pore   185
toxin   211
trafficking   15, 127, 170, 212

tranquilizer   182
trans-bilayer
  profile   81, 82
  structure   95
trans-double bond   45
trans-fatty acid   45
transfersome   200, 218
translational degrees of freedom   92
transmembrane helix   145
tri-acylglycerol (triglyceride)   25, 191, 194
triggering process   77, 164, 221
triterpenes   156
Triton X-100   124, 126
tryptophan   141
tubular myelin   115
two-photon fluorescence   121
type II diabetes   226

unbinding of membranes   61
undulation force   58, 61, 132, 133
unfolding of proteins   145
unsaturated lipid   23
  oxidation   194

vaccine   218
van der Waals force   5
venom   189
vesicle   37, 38, 54
  budding   70, 207
  budding transition   68
  COPII   172
  equilibrium shape   68
  fusion   131, 170
  in drug delivery   214
  multi-lamellar   37
  shape change   70
  shapes   70
  size   63
  trafficking   212
  transport   213
  transport at nerve cells   172
  uni-lamellar   63
viral vector   218
vision   168
visual system development   177
vitamin
  transport   156
  vitamin D   29

vitamin E   194

water   91
  -soluble protein   127
  boiling   94
  chemical potential   218
  chemical potential in skin   200
  diffusion across membrane   68, 82, 168
  diffusion along surface   42
  dipole moment   41
  hydrogen bonds   33, 34
  molecule   33
  surface tension   105
  the biological solvent   2, 33, 40

transport in proteins   159
wax   23, 193
wetting   41, 159
White, Steven   83
Wieland, Heinrich O.   149
Wieslander, Åke   51
Woese, Carl   202
Wortis, Michael   71

X-ray scattering   81, 110
  protein structure   169

yeast   9

zebra   176

Printing and Binding: Strauss GmbH, Mörlenbach